SpringerBriefs in Mathematical Physics

Volume 30

SpringerBriefs are characterized in general by their size (50-125 pages) and fast production time (2-3 months compared to 6 months for a monograph).
Briefs are available in print but are intended as a primarily electronic publication to be included in Springer's e-book package.

Typical works might include:

- An extended survey of a field
- A link between new research papers published in journal articles
- A presentation of core concepts that doctoral students must understand in order to make independent contributions
- Lecture notes making a specialist topic accessible for non-specialist readers.

SpringerBriefs in Mathematical Physics showcase, in a compact format, topics of current relevance in the field of mathematical physics. Published titles will encompass all areas of theoretical and mathematical physics. This series is intended for mathematicians, physicists, and other scientists, as well as doctoral students in related areas.

More information about this series at http://www.springer.com/series/11953

Hitoshi Murakami • Yoshiyuki Yokota

Volume Conjecture for Knots

 Springer

Hitoshi Murakami
Graduate School of Information Sciences
Tohoku University
Sendai, Japan

Yoshiyuki Yokota
Department of Mathematics and Information
Science
Tokyo Metropolitan University
Tokyo, Japan

ISSN 2197-1757   ISSN 2197-1765   (electronic)
SpringerBriefs in Mathematical Physics
ISBN 978-981-13-1149-9   ISBN 978-981-13-1150-5   (eBook)
https://doi.org/10.1007/978-981-13-1150-5

Library of Congress Control Number: 2018948365

This Springer imprint is published by Springer Nature, under the registered company Springer Nature Singapore Pte Ltd.
The registered company address is: 152 Beach Road, #21-01/04 Gateway East, Singapore 189721, Singapore

Preface

When V. Jones introduced his celebrated polynomial invariant, the Jones polynomial, in 1985, very few relations to topology were known. It originally came from operator algebra, and soon after another definition using the Kauffman bracket appeared. Thanks to Kauffman's approach, the Jones polynomial is now regarded as a topic that should be put in the first chapter of a textbook of knot theory. The Kauffman bracket uses planar diagrams combinatorially to define the Jones polynomial; it uses no homotopy or homology. Even though it proves a classical conjecture, Tait conjecture, saying that a reduced alternating diagram gives the minimum number of crossings for the corresponding knot, the Jones polynomial remained to be a mysterious invariant for knot theorists.

In 1989, E. Witten proposed a physical approach by using the so-called path integral. Roughly speaking, his "definition" of the Jones polynomial is to integrate the Chern–Simons action over "all" possible connections. Mathematically, it is an integral over an infinite dimensional space, and (so far) no rigorous definition is known. However, the idea is beautiful, and it also provides a way to construct a three-manifold invariant using the Jones polynomial.

On the other hand, in 1995, R. Kashaev used quantum dilogarithm to define a complex-valued knot invariant depending on an integer $N \geq 2$. He also conjectured that for the large asymptotic with respect to N, his invariant determines the hyperbolic volume of any *hyperbolic* knot.

In 1999, J. Murakami and the first author proved that Kashaev's invariant is nothing but a specialization of the colored Jones polynomial. More precisely, it coincides with the N-dimensional colored Jones polynomial evaluated at the N-th root of unity. They also proposed a conjecture generalizing Kashaev's one: the volume conjecture. It conjectures that for *any* knot the large N asymptotic of Kashaev's invariant gives the simplicial volume of the knot compliment. Here the simplicial volume is a generalization of the hyperbolic volume.

Soon after, Kashaev and O. Tirkkonen proved the volume conjecture for torus knots, whose simplicial volumes are known to be zero. The conjecture is also proved for the figure-eight knot, the simplest hyperbolic knot by T. Ekholm.

The volume conjecture fascinated not only knot theorists but also physicists.

The aim of this book is to study the volume conjecture from a mathematical viewpoint.

Chapter 1 contains preliminaries, describing basic facts about knots including the satellite construction of a knot, the torus decomposition of a knot complement, and braids. In Chap. 2 we describe how to construct topological invariants of a knot. In Sect. 2.1 a braid description is used to define the colored Jones polynomial and the Kashaev invariant. We also use a diagrammatic approach to them in Sect. 2.2. The volume conjecture is introduced in Chap. 3. It is proved for the cases of the figure-eight knot and a family of torus knots. In Chap. 4 we describe why we think that the volume conjecture is true. In Chap. 5 we prepare some facts about representations of the fundamental group of a knot complement to the Lie group $SL(2; \mathbb{C})$. We also define the Chern–Simons invariant and the Reidemeister torsion, both of which are associated with such a representation. In Chap. 6 the volume conjecture is generalized to a conjecture that is "twisted" by a representation.

We try to include as many examples as we can so that the readers can easily follow us.

Acknowledgements This work was supported by JSPS KAKENHI Grant Numbers JP17K05239 and JP15K04878.

Sendai, Japan   Hitoshi Murakami
Tokyo, Japan   Yoshiyuki Yokota
April, 2018

Contents

Acronyms

$\cong$	homeomorhic
α	Abelianization homomorphism
$\Lambda(z)$	Lobachevsky function
$\pi_1(K)$	knot group
$\psi_N(z)$	quantum dilogarigthm
$\mathbb{C}$	complex numbers
cs(M)	SO(3) Chern–Simons invariant of a hyperbolic three-manifold M
$\mathrm{CS}_{u,v}(\rho)$	SL(2; $\mathbb{C}$) Chern–Simons invariant of a representaion ρ
cv(M)	complex volume of a hyperbolic three-manifold M
D^n	n-dimensional disk
$\mathscr{E}$	figure-eight knot
$\mathbb{H}^3$	upper half model of the hyperbolic space
Int	interior
$J_N(K; q)$	N-dimensional colored Jones polynomial of a knot K
$\langle K \rangle_N$	Kashaev's invariant of a knot K
$\mathrm{Li}_2(z)$	dilogarithm function
lk	linking number
$\{m\}$	$q^{m/2} - q^{-m/2}$
$\{m\}!$	$\{m\}\{m-1\}\cdots\{2\}\{1\}$
$N(K)$	regular neighborhood of a knot K in S^3
$\mathbb{R}$	real numbers
$\mathbb{R}^n$	n-dimensional Euclidean space
$\mathrm{sl}_2(\mathbb{C})$	Lie algebra consisting of 2×2 matrices with trace 0
SL(2; $\mathbb{C}$)	Lie group consisting of 2×2 matrices with determinant 1
S^n	n-dimensional sphere
$\mathbb{T}^K_\mu(\rho)$	twisted Reidemeister torsion of a representation ρ
$T(p, q)$	torus knot of type (p, q)
Vol	simplicial volume
$\mathbb{Z}$	integers

Chapter 1
Preliminaries

Abstract In this chapter we describe fundamental definitions and theorems. For details, see for example Burde et al. (Knots, extended ed., De Gruyter studies in mathematics, vol 5. De Gruyter, Berlin, 2014. MR 3156509), Lickorish (An introduction to knot theory. Graduate texts in mathematics, vol 175. Springer, New York, 1997. MR 98f:57015), and Rolfsen (Knots and links. Mathematics lecture series, vol 7. Publish or Perish, Inc., Houston, 1990; Corrected reprint of the 1976 original. MR 1277811 (95c:57018)).

1.1 Knot

A *knot* is a circle smoothly embedded in the three-sphere S^3. Two knots are equivalent if and only if there exists a diffeomorphism of S^3 to itself taking one to the other. Usually we take orientation(s), of the circle and/or S^3, into account. See Fig. 1.1 for examples of knots (these pictures were drawn by Mathematica [89]).

It is often useful to consider $\mathbb{R}^3$ rather than S^3, regarding S^3 as the one-point compactification of $\mathbb{R}^3$, that is, $S^3 = \mathbb{R}^3 \cup \{\infty\}$.

If a knot is given in $\mathbb{R}^3$, then we can project it to the plane $\mathbb{R}^2 \subset \mathbb{R}^3$. We assume that the image does not have tangencies or multiple points except for double points. We draw the image on the plane so that at each double point the 'lower' one is broken as in Figs. 1.2 and 1.3. If a knot is oriented, we indicate it by an arrow (Fig. 1.4).

We call the image together with over/under information at each crossing a diagram of the knot (Figs. 1.3 and 1.4). If a diagram has no crossings, the corresponding knot is called the unknot (Fig. 1.1).

Of course there are infinitely many knot diagrams for a knot. However two knot diagrams of a knot can be transformed to each other by some simple 'moves' [74].

Definition 1.1 (Reidemeister moves) The following local moves are called Reidemeister moves I (Fig. 1.5), II (Fig. 1.6), and III (Fig. 1.7), respectively.

© The Author(s), under exclusive licence to Springer Nature Singapore Pte Ltd. 2018
H. Murakami, Y. Yokota, *Volume Conjecture for Knots*, SpringerBriefs in
Mathematical Physics 30, https://doi.org/10.1007/978-981-13-1150-5_1

Fig. 1.1 The unknot, the trefoil, and the figure-eight knot

Fig. 1.2 A crossing is indicated by breaking one of the lines

Fig. 1.3 A knot diagram

Fig. 1.4 An oriented knot

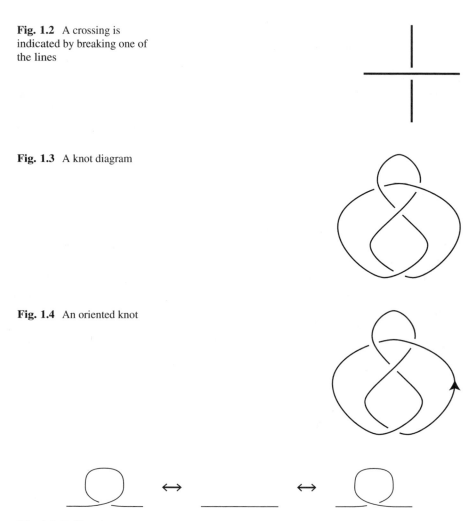

Fig. 1.5 Reidemeister move I

Fig. 1.6 Reidemeister move II

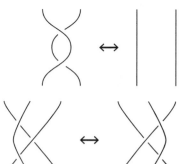

Fig. 1.7 Reidemeister move III

The following theorem is well known.

Theorem 1.1 (Reidemeister's theorem) *If two knot diagrams present equivalent knots, then they can be transformed to each other by a finite sequence of Reidemeister moves I, II, and III.*

For a proof see for example [14, 1.C].

1.2 Satellite

In this section we show several ways to construct knots from other knots.

See for example [76] for more details.

Definition 1.2 (satellite) Let C be a knot (in S^3) and P a knot in a solid torus $T \cong D^2 \times S^1$. If $e: T \to S^3$ is an embedding and the image $e(T)$ is a tubular neighborhood of C in S^3, then the image $e(P)$ is a knot (in S^3), called a satellite of C. We call C a companion and P a pattern of $e(P)$ (Fig. 1.8).

Note that even if C and $P \subset T$ are given, there are different ways to construct a satellite (see Fig. 1.9).

If there exists an embedded disk in T so that it intersects P with one point, then the satellite is called the connected sum of C and P, denoted by $C \sharp P$ (Fig. 1.10). It can be shown that $C \sharp P$ is uniquely determined.

Definition 1.3 (cable) If the pattern P is on the boundary of T, a satellite $e(P)$ is called a cable of C. Since a non-trivial closed curve on a torus, the boundary of T, is parametrized by a pair of coprime integers (p, q), we denote by $C^{(p,q)}$ the satellite of C with pattern P that travels p times along the knot and q times around

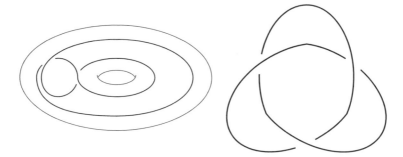

Fig. 1.8 Pattern P (left) and companion C (right)

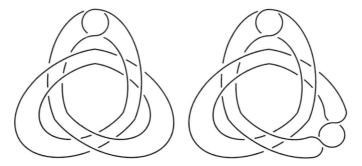

Fig. 1.9 Two satellites of C with pattern P given in Fig. 1.8

Fig. 1.10 Connected-sum of
the trefoil and the figure-eight
knot

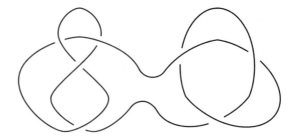

the knot.[1] Here a closed curve on a torus is called trivial if it bounds a disk in the torus. Note that we allow $(p, q) = (0, 1)$ and $(p, q) = (1, 0)$. In the former (latter, respectively) case it presents the *meridian* (*longitude*, respectively). We call $C^{(p,q)}$ the (p, q)-*cable* of the knot C. Figure 1.12 is the $(2, 3)$-cable of the trefoil, since it crosses the longitude three times in the positive direction.

---

[1] Precisely speaking, p counts how many times P intersects with the meridian (a circle on the torus that bounds a disk inside T) and q counts how many times P intersects with the longitude (a circle on the torus that is null-homologous outside T). The dotted circle in Fig. 1.11 shows the longitude of the trefoil. Observe that the linking number between the two circles is 0, since the dotted line goes under the solid line three times in the positive direction and three times in the negative direction.

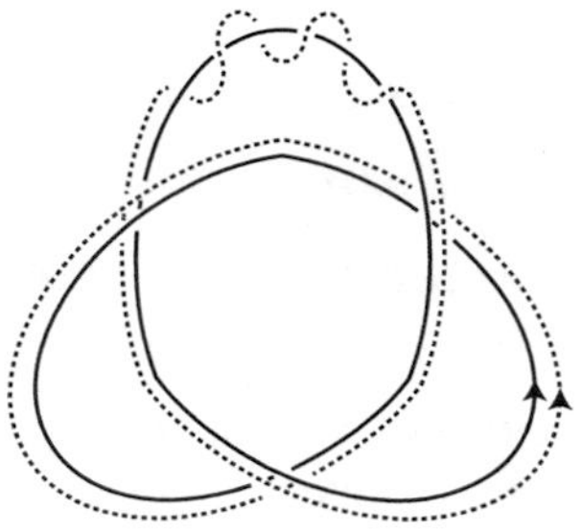

Fig. 1.11 A longitude of the trefoil

Fig. 1.12 (2, 3)-cable of the trefoil

Fig. 1.13 $T(3, 5)$ drawn by Mathematica

Fig. 1.14 $T(3, 5)$ on an unknotted torus, also drawn by Mathematica

Note that $C^{(1,q)}$ is equivalent to C.

Definition 1.4 (torus knot) For coprime integers (p, q) with $p > 1$, the torus knot of type (p, q), denoted by $T(p, q)$, is the satellite knot $U^{(p,q)}$, where U is the unknot. The knot $T(2, 3)$ is the trefoil knot and $T(3, 5)$ is indicated in Fig. 1.13. So a torus knot is a knot that can be drawn on an unknotted torus. See Fig. 1.14.

Definition 1.5 (simple knot) If a knot K is not a satellite of any knot other than the unknot, then K is called simple.

The following theorem is well known.

Theorem 1.2 (W. Thurston) *A simple knot is either a torus knot or a* hyperbolic knot.

Here a knot is called hyperbolic if it possesses a complete hyperbolic structure with finite volume.

If K is not simple, there exists an incompressible[2] and non-boundary-parallel torus[3] in the complement $S^3 \setminus K$. Considering a 'maximal' set of such tori we can decompose the knot complement into several pieces in a unique way.

Theorem 1.3 (Jaco–Shalen–Johannson decomposition [35, 36]) *Let K be a knot in S^3. There exists a maximal set of incompressible tori in $S^3 \setminus K$. Here 'maximal' means that there are no pair of parallel tori nor boundary-parallel torus.*

This decomposition is called the Jaco–Shalen–Johannson decomposition or torus decomposition.

By using this decomposition we can define the simplicial volume of a knot.

Definition 1.6 (simplicial volume) Let K be a knot and consider its Jaco–Shalen–Johannson decomposition $(S^3 \setminus K) \setminus \mathscr{T}$, where $\mathscr{T}$ is a maximal set of incompressible tori. Then each of its connected components is either

- hyperbolic, that is, it possesses a complete hyperbolic structure with finite volume, or
- Seifert fibered, that is, it is a circle bundle over a surface with singularities.

The *simplicial volume* $\mathrm{Vol}(S^3 \setminus K)$ is defined to be the sum of the *hyperbolic volumes* of the hyperbolic pieces.

Remark 1.1 The volume in Definition 1.6 coincides with the Gromov norm up to multiplication by a constant [28, 80]. Note that the Gromov norm of a Seifert fibered space is 0.

Remark 1.2 For a prime,[4] closed, oriented three-manifold, Thurston's geometrization conjecture (now a theorem by G. Perelman) says that after suitable decomposition like the JSJ decomposition above, each piece possesses one of the following eight geometries: (1) hyperbolic, (2) spherical, (3) Euclidean, (4) $\mathbb{R} \times S^1$, (5) $\mathbb{R} \times \mathbb{H}^2$, (6) Nil, (7) Sol, (8) $\widetilde{\mathrm{SL}}(2; \mathbb{R})$. See [57, 81] for details.

---

[2]A surface S in a three-manifold M is incompressible if the inclusion $\pi_1(S) \to \pi_1(M)$ is injective.

[3]Two tori are parallel if they bound a thickened torus ($S^1 \times S^1 \times [0, 1]$) and a torus in a knot complement is called boundary-parallel if it is parallel to the boundary of a tubular neighborhood of the knot in S^3.

[4]A closed three-manifold is called prime if it cannot be a connected-sum of two three-manifolds, none of which is the three-sphere.

Fig. 1.15 The (2, 1)-cable of
the figure-eight knot

Fig. 1.16 Figure-eight knot

Example 1.1 (hyperbolic knot) Since a hyperbolic knot K is simple, the JSJ decomposition of $S^3 \setminus K$ is just itself. So the simplicial volume of K is its hyperbolic volume.

Example 1.2 (torus knot) Since a torus knot $T(p, q)$ is simple, the JSJ decomposition of $S^3 \setminus T(p, q)$ is itself as in the previous case. On the other hand its simplicial volume is 0.

Example 1.3 Let $\mathscr{E}^{(2,1)}$ denote the (2, 1)-cable of the *figure-eight knot* $\mathscr{E}$. It is depicted in Fig. 1.15. Compare it with the figure-eight knot (Fig. 1.16).

Figure 1.17 shows a torus embedded in the knot complement $S^3 \setminus \mathscr{E}^{(2,1)}$. If we remove the torus from $S^3 \setminus \mathscr{E}^{(2,1)}$, then the complement is decomposed into the figure-eight knot complement $S^3 \setminus \mathscr{E}$ (the left part of the right hand side) and a solid torus ($D^2 \times S^1$) minus a knot going around it twice (the right part of the right hand side) as in (1.1).

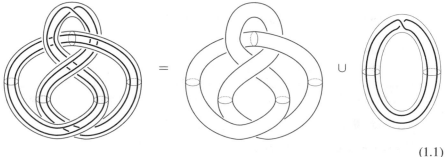

$$\tag{1.1}$$

It is known that $S^3 \setminus \mathscr{E}$ is hyperbolic and that the other piece is Seifert fibered.

Fig. 1.17 The complement
of the (2, 1)-cable of the
figure-eight knot

Fig. 1.18 A 3-braid

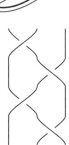

Therefore Vol $\left(S^3 \setminus \mathscr{E}^{(2,1)}\right) = \mathrm{V}(S^3 \setminus \mathscr{E})$, which equals $6\Lambda(\pi/3) = 2.02988\dots$
because it is well known that $S^3 \setminus \mathscr{E}$ can be decomposed into two ideal regular
tetrahedra. Here we use the *Lobachevsky function*:

$$\Lambda(\theta) := -\int_0^\theta \log|2\sin x| \, dx. \tag{1.2}$$

It is well known that the volume of an ideal hyperbolic tetrahedron with dihedral
angles α, β, γ is $\Lambda(\alpha) + \Lambda(\beta) + \Lambda(\gamma)$ (see for example [60] for details).

1.3   Braid

In this section we introduce fundamental facts about braids. See [12, 42] for more
details.

Let I the closed interval $[0, 1]$. An *n-braid* consists of n strings in I^3 such that
each string connecting a point $I^2 \times \{1\}$ and a point in $I^2 \times \{0\}$ monotonically. We
assume that the i-th string connects $(i/(n + 1), 1/2, 1)$ and $(\tau(i)/(n + 1), 1/2, 0)$
$(i = 1, 2, \dots, n)$, where τ is an element in the symmetric group with n letters.
Figure 1.18 is a three-braid with $\tau = \begin{pmatrix} 1 & 2 & 3 \\ 2 & 3 & 1 \end{pmatrix}$. Two braids are equivalent if they are
isotopic to each other fixing the endpoints.

Fig. 1.19 The closure of Fig. 1.18

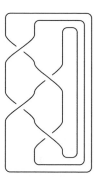

Fig. 1.20 The i-th generator σ_i

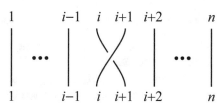

Given a braid, one can construct a knot (or a link, several circles in S^3) by closing it.

Definition 1.7 (closure of a braid) Let β be a braid. Then its closure $\hat{\beta}$ is obtained by connecting $(i/(n+1), 1/2, 0)$ and $(i/(n+1), 1/2, 1)$ as in Fig. 1.19.

Any knot can be presented as the closure of a braid.

Theorem 1.4 (Alexander's theorem [1]) *Any knot is equivalent to the closure of a braid.*

For example the knot in Fig. 1.16 is equivalent to the closure of the braid shown in Fig. 1.18 (see Figs. 1.18 and 1.19).

The set of all the n-braids forms a group in the following way. The product of two n-braids β_1 and β_2 is defined by putting β_1 on β_2, and shrink them vertically so that they fit in I^3. Then the braid consisting of n straight strings is the identity and the inverse of a braid is given by reflecting it vertically (put a mirror horizontally). Denote this group of n-braids by B_n.

It is know that the group B_n has the following presentation.

Theorem 1.5 (Artin [8]) *Let σ_i $(i = 1, 2, \ldots, n-1)$ be the n-braid depicted in Fig. 1.20. The braid group B_n has the following presentation.*

$$B_n = \langle \sigma_1, \sigma_2, \ldots, \sigma_{n-1} \mid \sigma_i \sigma_j = \sigma_j \sigma_i \quad (|i-j| > 1), \sigma_i \sigma_{i+1} \sigma_i = \sigma_{i+1} \sigma_i \sigma_{i+1}$$

$$(i=1,2,\ldots,n\text{-}2)\rangle. \tag{1.3}$$

Fig. 1.21 The braid relation
$\sigma_i\sigma_{i+1}\sigma_i = \sigma_{i+1}\sigma_i\sigma_{i+1}$

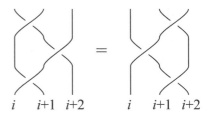

$i \quad i+1 \quad i+2 \qquad i \quad i+1 \quad i+2$

For example the braid in Fig. 1.18 is presented as $\sigma_1\sigma_2^{-1}\sigma_1\sigma_2^{-1}$. See Fig. 1.21 for the *braid relation* $\sigma_i\sigma_{i+1}\sigma_i = \sigma_{i+1}\sigma_i\sigma_{i+1}$.

Remark 1.3 In [12], σ_i is exchanged for σ_i^{-1}.

Note that the braid relation corresponds to the Reidemeister move III and $\sigma_i\sigma_i^{-1} = 1$ is the Reidemeister move II.

Two braids $\beta_1 \in B_n$ and $\beta_2 \in B_m$ may give equivalent knots (or links) as closures. Two such braids are related by a sequence of *Markov moves*.

Theorem 1.6 (Markov's theorem [54]) *Let $\beta_1 \in B_n$ and $\beta_2 \in B_m$ be braids, and $\hat{\beta}_1$ and $\hat{\beta}_2$ be their closures respectively. If $\hat{\beta}_1$ and $\hat{\beta}_2$ are equivalent then they are transformed to each other by a sequence of the following two moves (Markov moves).*

- *conjugation:* $\alpha\beta \Leftrightarrow \beta\alpha$,

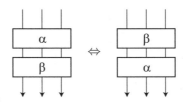

- *(de)stabilization:* $\beta \in B_n \Leftrightarrow \beta\sigma_n^{\pm 1} \in B_{n+1}$.

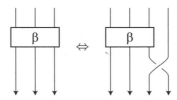

Chapter 2
R-Matrix, the Colored Jones Polynomial, and the Kashaev Invariant

Abstract In this chapter we give definitions of the colored Jones polynomial. To do that we use a braid presentation and a knot diagram. Kashaev's invariant is obtained as a specialization of the colored Jones polynomial.

2.1 A Link Invariant Derived from a Yang–Baxter Operator

2.1.1 Yang–Baxter Operator

Let V be an N-dimensional vector space over $\mathbb{C}$. For homomorphisms $R \colon V \otimes V \to V \otimes V$, $\mu \colon V \to V$ and non-zero complex numbers a, b, the quadruple (R, μ, a, b) is called an *enhanced Yang–Baxter operator* [83] if the following three equalities hold:

$$(R \otimes \mathrm{Id}_V)(\mathrm{Id}_V \otimes R)(R \otimes \mathrm{Id}_V) = (\mathrm{Id}_V \otimes R)(R \otimes \mathrm{Id}_V)(\mathrm{Id}_V \otimes R), \tag{2.1}$$

$$R(\mu \otimes \mu) = (\mu \otimes \mu)R, \tag{2.2}$$

$$\mathrm{Tr}_2(R^{\pm}(\mathrm{Id}_V \otimes \mu)) = a^{\pm 1} b \,\mathrm{Id}_V . \tag{2.3}$$

Here Id_V is the identity on V and $\mathrm{Tr}_k \colon \mathrm{End}(V^{\otimes k}) \to \mathrm{End}(V^{\otimes(k-1)})$ is the operator trace defined as

$$\mathrm{Tr}_k(f)(e_{i_1} \otimes e_{i_2} \otimes \cdots \otimes e_{i_{k-1}}) := \sum_{j_1, j_2, \ldots, j_{k-1}, j=0}^{N-1} f_{i_1, i_2, \ldots, i_{k-1}, j}^{j_1, j_2, \ldots, j_{k-1}, j} \, (e_{j_1} \otimes e_{j_2} \otimes \cdots \otimes e_{j_{k-1}} \otimes e_j),$$

where $\{e_0, e_1, \ldots, e_{N-1}\}$ is a basis of V and the $f_{i_1, i_2, \ldots, i_{k-1}, j}^{j_1, j_2, \ldots, j_{k-1}, j}$ are defined as

$$f(e_{i_1} \otimes e_{i_2} \otimes \cdots \otimes e_{i_k}) = \sum_{j_1, j_2, \ldots, j_{k-1}, j_k=0}^{N-1} f_{i_1, i_2, \ldots, i_k}^{j_1, j_2, \ldots, j_k} \, (e_{j_1} \otimes e_{j_2} \otimes \cdots \otimes e_{j_k}).$$

© The Author(s), under exclusive licence to Springer Nature Singapore Pte Ltd. 2018   11
H. Murakami, Y. Yokota, *Volume Conjecture for Knots*, SpringerBriefs in
Mathematical Physics 30, https://doi.org/10.1007/978-981-13-1150-5_2

Note that the definition of Tr_k does not depend on the choice of bases and that Tr_1 is the usual trace on matrices.

Equation (2.1) is called the *Yang–Baxter equation* [10, 11, 91] and R is called the *R-matrix*.

Given a knot K, let β be an n-braid such that its closure $\hat{\beta}$ is equivalent to K. If β is presented as a product of generators given in (1.3) and their inverses, then we replace each $\sigma_i^{\pm 1}$ with

$$\mathrm{Id}_V^{\otimes i-1} \otimes R^{\pm 1} \otimes \mathrm{Id}_V^{\otimes n-i-1} \colon V^{\otimes n} \to V^{\otimes n},$$

where $\mathrm{Id}_V^{\otimes j}$ means the j-fold tensor of Id_V. Then we have a homomorphism $\Phi(\beta)$ from $V^{\otimes n}$ to itself. For example the braid given in Fig. 1.18 defines a homomorphism $(R \otimes \mathrm{Id}_V)(\mathrm{Id}_V \otimes R^{-1})(R \otimes \mathrm{Id}_V)(\mathrm{Id}_V \otimes R^{-1}) \colon V \otimes V \otimes V \to V \otimes V \otimes V$ as shown in Fig. 2.1. Note that the homomorphism is from the top to the bottom.

By taking $\mathrm{Tr}_n, \mathrm{Tr}_{n-1}, \ldots, \mathrm{Tr}_1$ of $\Phi(\beta)$ successively we can define a knot invariant.

Definition 2.1 ([83]) Let (R, μ, a, b) be an enhanced Yang–Baxter operator. For a knot K presented by the closure of an n-braid β, define

$$T_{(R,\mu,a,b)}(K) := a^{-w(\beta)} b^{-n} \, \mathrm{Tr}_1(\mathrm{Tr}_2(\cdots(\mathrm{Tr}_n(\Phi(\beta)\mu^{\otimes n})))) \in \mathbb{C}. \qquad (2.4)$$

Here $w(\beta)$ is the sum of the exponents in β. Then this scalar is an invariant of links, that is, the right hand side does not depend on braids that present the knot K.

For example, if $\beta = \sigma_1 \sigma_2^{-1} \sigma_1 \sigma_2^{-1}$ as in Fig. 2.1, the right hand side of (2.4) can be depicted in Fig. 2.2. Note that closing a string corresponds to taking Tr_k. To prove the well-definedness, one needs to check that $T_{(R,\mu,a,b)}$ is invariant under the braid relation, the conjugation, and (de)stabilization from Theorem 1.6. The invariance under the braid relation follows from (2.1):

Fig. 2.1 The homomorphism defined by Fig. 1.18

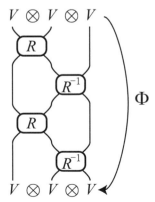

Fig. 2.2 The invariant
$T_{(B,\mu,a,b)}$ defined by Fig. 1.18

$$a^{-w(\beta)}b^{-n}\times$$

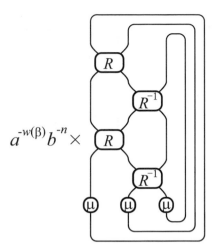

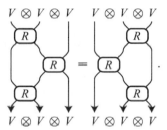

The invariance under the conjugation follows from (2.2):

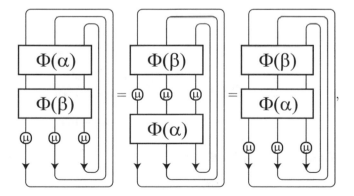

where the first equality holds since Tr_k is invariant under conjugation. The invariance under (de)stabilization follows from (2.3):

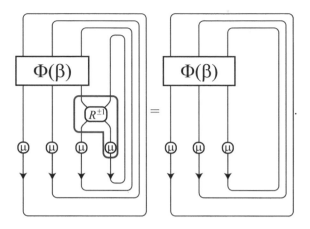

2.1.2   Colored Jones Polynomial

We give an enhanced Yang–Baxter operator for each integer $N \geq 2$ that gives the N-dimensional *colored Jones polynomial*.

Let V be the N-dimensional complex vector space $\mathbb{C}^N$. We define $R \colon V \otimes V \to V \otimes V$ and $\mu \colon V \to V$ as follows [45, 46]. Let $\{e_0, e_1, \ldots, e_{N-1}\}$ be the standard basis of V. For a complex parameter q, we define $\{m\} := q^{m/2} - q^{-m/2}$ and $\{m\}! := \{m\}\{m-1\} \cdots \{2\}\{1\}$. We put

$$R(e_k \otimes e_l) := \sum_{i,j=0}^{N-1} R_{kl}^{ij} e_i \otimes e_j$$

with

$$R_{kl}^{ij} := \sum_{m=0}^{\min(N-1-i,j)} \delta_{l,i+m} \delta_{k,j-m} \frac{\{l\}!\{N-1-k\}!}{\{i\}!\{m\}!\{N-1-j\}!} \tag{2.5}$$
$$\times \, q^{(i-(N-1)/2)(j-(N-1)/2)-m(i-j)/2-m(m+1)/4}.$$

Here $\delta_{i,j}$ is Kronecker's delta. We also put

$$\mu(e_j) := \sum_{i=0}^{N-1} \mu_j^i e_i$$

with

$$\mu_j^i := \delta_{i,j} q^{(2i-N+1)/2}.$$

Then it can be proved that $(R, \mu, q^{(N^2-1)/4}, 1)$ is an enhanced Yang–Baxter operator.

Now for a knot K, we define the N-dimensional colored Jones polynomial $J_N(L; q)$ as follows:

$$J_N(K; q) := \frac{\{1\}}{\{N\}} T_{(R,\mu,q^{(N^2-1)/4},1)}(K). \tag{2.6}$$

Then this is a knot invariant. Note that we normalized it so that the colored Jones polynomial of the unknot U is 1 because

$$T_{(R,\mu,q^{(N^2-1)/4},1)}(U) = \mathrm{Tr}_1(\mu) = \sum_{i=0}^{N-1} q^{(2i-N+1)/2} = \frac{\{N\}}{\{1\}}.$$

When $N = 2$, the matrix R is presented as follows with respect to the basis $\{e_0 \otimes e_0, e_0 \otimes e_1, e_1 \otimes e_0, e_1 \otimes e_1\}$:

$$R = \begin{pmatrix} q^{1/4} & 0 & 0 & 0 \\ 0 & q^{1/4} - q^{-3/4} & q^{-1/4} & 0 \\ 0 & q^{-1/4} & 0 & 0 \\ 0 & 0 & 0 & q^{1/4} \end{pmatrix}.$$

The matrix μ is given as follows with respect to the basis $\{e_0, e_1\}$:

$$\mu = \begin{pmatrix} q^{-1/2} & 0 \\ 0 & q^{1/2} \end{pmatrix}.$$

Therefore we have the following equality:

$$q^{1/4} R - q^{-1/4} R^{-1} = (q^{1/2} - q^{-1/2}) \, \mathrm{Id}_V \otimes \mathrm{Id}_V . \tag{2.7}$$

Taking $w(\beta)$ into account we have the following skein relation:

$$q J_2\left(\reflectbox{$\diagup$}\kern-1em\diagdown\,; q\right) - q^{-1} J_2\left(\diagdown\kern-1em\diagup\,; q\right) = (q^{1/2} - q^{-1/2}) J_2\left(\,\rangle\ \langle\,; q\right).$$

Therefore the 2-dimensional colored Jones polynomial coincides with the original Jones polynomial $V(K; q)$ [37].

2.1.3   Kashaev's R-Matrix

Kashaev introduced the following R-matrix [38].

Put $(x)_n = \prod_{i=1}^{n}(1 - x^i)$ for $n \geq 0$. Define $\theta : \mathbb{Z} \to \{0, 1\}$ by

$$\theta(n) = \begin{cases} 1 & \text{if } N > n \geq 0, \\ 0 & \text{otherwise.} \end{cases}$$

For an integer x, we denote by $\text{res}(x) \in \{0, 1, 2, \ldots, N-1\}$ the residue modulo N. Now *Kashaev's R-matrix* R_K is given by

$$(R_\text{K})_{ab}^{cd}$$

$$= N\xi^{1+c-b+(a-d)(c-b)} \frac{\theta(\text{res}(b-a-1) + \text{res}(c-d))\theta(\text{res}(a-c) + \text{res}(d-b))}{(\xi)_{\text{res}(b-a-1)}(\xi^{-1})_{\text{res}(a-c)}(\xi)_{\text{res}(c-d)}(\xi^{-1})_{\text{res}(d-b)}},$$

where $\xi := \exp(2\pi\sqrt{-1}/N)$. Putting $(\mu_\text{K})_j^i := -\xi^{1/2}\delta_{i,j+1}$, the quadruple $(R_\text{K}, \mu_\text{K}, -\xi^{1/2}, 1)$ is also an enhanced Yang–Baxter operator. In [66], it is proved that replacing q with ξ, $(R, \mu, q^{(N^2-1)/4}, 1)$ defines the same knot invariant with the one defined by $(R_\text{K}, \mu_\text{K}, -\xi^{1/2}, 1)$.

2.1.4   Example of Calculation

As an example, we will calculate the colored Jones polynomial of the figure-eight knot $\mathscr{E}$. Put $\beta := \sigma_1\sigma_2^{-1}\sigma_1\sigma_2^{-1}$. Then its closure is equivalent to $\mathscr{E}$ (Fig. 1.19). We will calculate $J_N(\mathscr{E}; q)$ by using the enhanced Yang-Baxter operator $(R, \mu, q^{(N-2-1)/4}, 1)$.

By Definition 2.1 and (2.6), we have

$$J_N(\mathscr{E}; q) = \frac{\{1\}}{\{N\}} \text{Tr}_1(\text{Tr}_2(\text{Tr}_3(\Phi(\beta)\mu^{\otimes 3})))$$

since $w(\beta) = 0$. However it is easier to calculate $\text{Tr}_2(\text{Tr}_3(\Phi(\beta)(\text{Id}_V \otimes \mu \otimes \mu))) \in \text{End}(V)$, which coincides with $S \times \text{Id}_V$ for a scalar S by Schur's lemma (see [45, Lemma 3.9] for a proof). See Fig. 2.3.

Then since $\text{Tr}_1(\text{Tr}_2(\text{Tr}_3(\Phi(\beta)\mu^{\otimes 3}))) = S \times \text{Tr}_1(\mu) = \frac{\{N\}}{\{1\}}S$, we have $J_N(L; q) = S$.

More explicitly, the scalar S becomes

$$\sum_{b,c,d,e,f,g,h} R_{c,d}^{a,b} (R^{-1})_{f,g}^{d,e} R_{a,h}^{c,f} (R^{-1})_{b,e}^{h,g} \mu_b^b \mu_e^e, \tag{2.8}$$

Fig. 2.3 We close all the
strings except for the first one

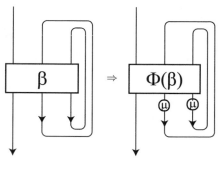

Fig. 2.4 Labels $a, b, \ldots, h$
assigned to arcs

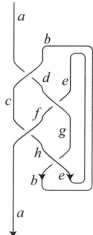

which does not depend on a. This can be depicted in Fig. 2.4. Here we associate the
R-matrix or its inverse to each crossing as follows.

$$\begin{array}{cc} i \quad j \\ \diagdown\!\diagup \end{array} \Rightarrow R^{ij}_{kl} \ , \qquad \begin{array}{cc} i \quad j \\ \diagup\!\diagdown \end{array} \Rightarrow (R^{-1})^{ij}_{kl}$$
$$\begin{array}{cc} k \quad l \end{array} \qquad\qquad \begin{array}{cc} k \quad l \end{array}$$

Note that the inverse of the R-matrix is given by

$$(R^{-1})^{ij}_{kl} = \sum_{m=0}^{\min(N-1-i,j)} \delta_{l,i-m}\delta_{k,j+m} \frac{\{k\}!\{N-1-l\}!}{\{j\}!\{m\}!\{N-1-i\}!}$$

$$\times (-1)^m q^{-\left(i-(N-1)/2\right)\left(j-(N-1)/2\right)-m(i-j)/2+m(m+1)/4}. \tag{2.9}$$

Since (2.8) does not depend on a, we put $a := N - 1$ (Fig. 2.5).

Fig. 2.5 Put $a := N - 1$

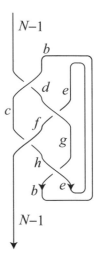

From Kronecker's deltas in (2.5) and (2.9), we ignore labelings which do not satisfy the following rules:

(+). At a positive crossing, the top-left label is less than or equal to the bottom-right label, the top-right label is greater than or equal to the bottom-left label. Moreover the sum of the top two labels equals the sum of the bottom two labels (see (2.5)).

$$: i+j = k+l, l \geq i, k \leq j,$$

(−). At a negative crossing, the top-left label is greater than or equal to the bottom-right label, the top-right label is less than or equal to the bottom-left label. Moreover the sum of the top two labels equals the sum of the bottom two labels (see (2.9)).

$$: i+j = k+l, l \leq i, k \geq j.$$

Now look at Fig. 2.5. From Rule (+), we have $d = N - 1$ and $c = b$ (Fig. 2.6). Applying Rule (−) to the second crossing, we have $N - 1 + e = f + g$ and so $f = N - 1 + e - g$ (Fig. 2.7). Applying Rule (+) to the third crossing, we have $N - 1 + e - g + b = N - 1 + h$ and so $h = e - g + b$ (Fig. 2.8). From the inequalities

Fig. 2.6 $d = N - 1, c = b$

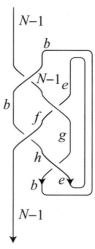

Fig. 2.7 $f = N - 1 + e - g$

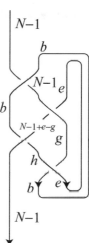

of Rule $(+)$, we have $N - 1 + e - g \geq N - 1$ and $b \leq e - g + b$. Therefore we have $g = e$ (Fig. 2.9) and (2.8) becomes

$$\sum_{b \geq e} R_{b,N-1}^{N-1,b} \, (R^{-1})_{N-1,e}^{N-1,e} \, R_{N-1,b}^{b,N-1} \, (R^{-1})_{b,e}^{b,e} \, \mu_b^b \, \mu_e^e$$

$$= \sum_{b \geq e} (-1)^{N-1+b} \frac{\{N-1\}!\{b\}!\{N-1-e\}!}{(\{e\}!)^2 \{b-e\}!\{N-1-b\}!} \tag{2.10}$$

$$\times \, q^{(-b-b^2-2be-2j^2+3N+6Nb+2Ne-3N^2)/4},$$

where we use Rule $(-)$ at the fourth crossing to get the inequality $b \geq e$.

Fig. 2.8 $h = e - g + b$

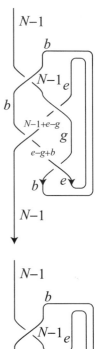

Fig. 2.9 $g = e$

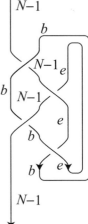

It is sometimes useful to regard a knot as the closure of a $(1, 1)$-tangle[1] as shown in Fig. 2.10.

In this case we need to follow Rules $(+)$, $(-)$ and

($\smile$) Put μ at each local minimum where the arc goes from left to right,
($\frown$) Put μ^{-1} at each local maximum where the arc goes from left to right.

See [45, Theorem 3.6] for details.

---

[1] A properly embedded string in I^3 with one endpoint in $I^2 \times \{0\}$ and the other in $I^2 \times \{1\}$. Note that we allow maxima or minima.

Fig. 2.10 The figure-eight knot can also regarded as the closure of a (1, 1)-tangle

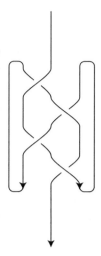

Fig. 2.11 (1, 1)-tangle with labels

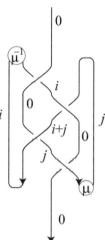

If we put 0 at the top and the bottom, the other labelings become as depicted in Fig. 2.11. So we have

$$J_N(\mathscr{E}; q)$$

$$= \sum_{\substack{0 \leq i \leq N-1, 0 \leq j \leq N-1 \\ 0 \leq i+j \leq N-1}} R^{i,0}_{0,i} \, (R^{-1})^{i,j}_{i+j,0} \, R^{0,i+j}_{i,j} \, (R^{-1})^{j,0}_{0,j} \, (\mu^{-1})^i_i \, \mu^j_j$$

$$= \sum_{\substack{0 \leq i \leq N-1, 0 \leq j \leq N-1 \\ 0 \leq i+j \leq N-1}} (-1)^i \frac{\{i+j\}! \{N-1\}!}{\{i\}! \{j\}! \{N-1-i-j\}!} q^{-(N-1)i/2 + (N-1)j/2 - i^2/4 + j^2/4 - 3i/4 + 3j/4}.$$

Putting $k := i + j$, this becomes

$$J_N(\mathscr{E}; q) = \sum_{k=0}^{N-1} \frac{\{N-1\}!}{\{N-1-k\}!} q^{k^2/4+Nk/2+k/4} \left(\sum_{i=0}^{k} (-1)^i \frac{\{k\}!}{\{i\}!\{k-i\}!} q^{-Ni-ik/2-i/2} \right).$$

Using the formula (see [66, Lemma 3.2])

$$\sum_{i=0}^{k} (-1)^i q^{li/2} \frac{\{k\}!}{\{i\}!\{k-i\}!} = \prod_{g=1}^{k} (1 - q^{(l+k+1)/2-g}),$$

we have the following formula with only one summand, which is originally due to K. Habiro [30] and T. Lê.

$$J_N(\mathscr{E}; q) = \frac{1}{\{N\}} \sum_{k=0}^{N-1} \frac{\{N+k\}!}{\{N-1-k\}!}. \tag{2.11}$$

2.2   Colored Jones Polynomial via the Kauffman Bracket

2.2.1   Kauffman Bracket

There is another way to calculate the colored Jones polynomial. Given an unoriented link diagram $|D|$, one can define the *Kauffman bracket* $\langle |D| \rangle$ by using the following two axioms.

$$\langle U^c \rangle = (-A^2 - A^{-2})^c,$$

where U^c is the trivial c component link diagram [44]. The 2-dimensional colored Jones polynomial $J_2(K; t)$ of a knot K with a diagram D is defined as

$$\left. \frac{(-A^3)^{-w(D)} \langle |D| \rangle}{-A^2 - A^{-2}} \right|_{q:=A^4},$$

where $|D|$ is the unoriented diagram obtained from D by forgetting the orientation, and $w(D)$ is the writhe of D (the sum of the signs of D; positive for ✘ ✘ and negative for ✘ ✘).

Now define the *Jones–Wenzl idempotent* [87] by the following recurrence relation.

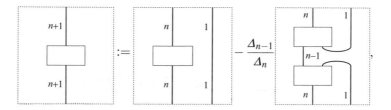

where an integer beside an arc is the number of parallel copies of the arc and $\Delta_n :=$ $(-1)^n \frac{A^{2(n+1)}-A^{-2(n+1)}}{A^2-A^{-2}}$. The N-colored Jones polynomial is defined as

$$\frac{\left((-1)^{N-1}A^{N^2-1}\right)^{-w(D)} \left\langle \vcenter{\hbox{\includegraphics{}}} \right\rangle}{\Delta_{N-1}} \Bigg|_{q:=A^4}.$$

See, for example, [56] or [53, Chapter 14] for more details.

For actual calculation the following formulas are useful:

$$\left\langle \vcenter{\hbox{\includegraphics{}}} \right\rangle = \sum_{c=0}^{d} \frac{\Delta_{2c}}{\theta(b,b,2c)} \left\langle \vcenter{\hbox{\includegraphics{}}} \right\rangle, \tag{2.12}$$

$$\vcenter{\hbox{\includegraphics{}}} = (-1)^{(a+b-c)/2} A^{a+b-c+(a^2+b^2-c^2)/2} \vcenter{\hbox{\includegraphics{}}}, \tag{2.13}$$

where $\theta(a, b, c)$ is defined as

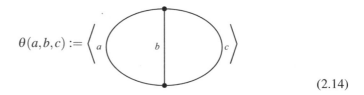

$$\theta(a,b,c) := \qquad\qquad\qquad\qquad\qquad\qquad\qquad (2.14)$$

and a trivalent vertex means

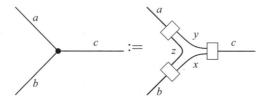

with $a = y + z$, $b = z + x$ and $c = x + y$. A precise formula for $\theta(a, b, c)$ can be found in [56].

2.2.2 Example of Calculation

In this subsection we calculate the colored Jones polynomial of the torus knot $T(2, 2a + 1)$ $(a > 0)$ (Fig. 2.12) by using the Kauffman bracket.

We first calculate the Kauffman bracket of the diagram that is obtained by replacing the knot diagram in Fig. 2.12 with the Jones–Wenzl idempotent. We have

Fig. 2.12 Torus knot
$T(2, 2a + 1)$

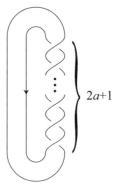

$2a+1$

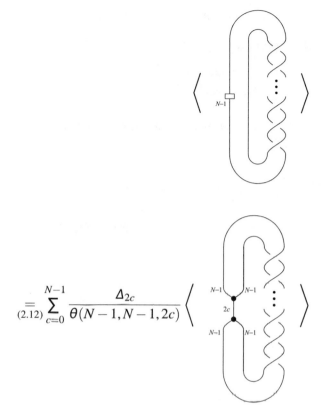

$$\underset{(2.12)}{=} \sum_{c=0}^{N-1} \frac{\Delta_{2c}}{\theta(N-1,N-1,2c)} \left\langle \quad \right\rangle$$

$$\underset{(2.13)}{=} \sum_{c=0}^{N-1} \frac{\Delta_{2c}}{\theta(N-1,N-1,2c)} \left((-1)^{c-N+1} A^{-2(N-1)+2c+2c^2-(N-1)^2} \right)^{2a+1}$$

$$\times \left\langle \quad \right\rangle$$

$$(2.15)$$

Therefore the colored Jones polynomial equals

$$J_N(T(2, 2a + 1); q)$$

$$=\frac{(-1)^{N-1}q^{-(2a+1)(N^2-1)/4}}{q^{N/2} - q^{-N/2}}$$

$$\times \sum_{c=0}^{N-1}(-1)^c q^{(2a+1)(2c^2+2c-N^2+1)/4}\left(q^{(2c+1)/2} - q^{-(2c+1)/2}\right)$$

$$=\frac{(-1)^{N-1}q^{-(2a+1)(N^2-1)/2}}{q^{N/2} - q^{-N/2}}\sum_{c=0}^{N-1}(-1)^c q^{(2a+1)(c^2+c)/2}\left(q^{(2c+1)/2} - q^{-(2c+1)/2}\right).$$

$$(2.16)$$

A formula for a general torus knot can be found in [61, 77].

Chapter 3
Volume Conjecture

Abstract In Kashaev (Lett Math Phys 39(3):269–275, 1997. MR 1434238), Kashaev proposed a conjecture that his invariant $\langle K \rangle_N$ defined in Kashaev (Mod Phys Lett A 10(19):1409–1418, 1995. MR 1341338) would grow exponentially with respect to N and that its growth rate would give the hyperbolic volume of the complement of a hyperbolic knot. If we replace the parameter q in the colored Jones polynomial with $\exp(2\pi\sqrt{-1}/N)$, we can regard it as a function of a natural number $N \geq 2$. In Murakami and Murakami (Acta Math 186(1):85–104, 2001. MR 1828373), J. Murakami and the first author proved that this coincides with Kashaev's invariant. The volume conjecture states that this function would grow exponentially with respect to N and its growth rate would give the simplicial volume of the knot complement. In this section we describe the volume conjecture and give proofs for the figure-eight knot and for the torus knot $T(2, 2a + 1)$.

3.1 Volume Conjecture

Let $\langle K \rangle_N$ be the link invariant defined by using Kashaev's enhanced Yang–Baxter operator $(R_{\mathrm{K}}, \mu_{\mathrm{K}}, -\xi^{1/2}, 1)$ for a knot K (see Sect. 2.1.3) [38]. In [39] Kashaev conjectured that the following equality holds for any hyperbolic knot,

$$\lim_{N \to \infty} \frac{1}{N} \log |\langle K \rangle_N| = \frac{\mathrm{V}\left(S^3 \setminus K\right)}{2\pi},$$

where $\mathrm{V}(S^3 \setminus K)$ denotes the hyperbolic volume. J. Murakami and the first author proved in [66] that *Kashaev's invariant* $\langle K \rangle_N$ coincides with the colored Jones polynomial $J_N(K; q)$ evaluated at $q = \exp(2\pi\sqrt{-1}/N)$. We also generalized his conjecture for any knot.

H. Murakami, Y. Yokota, *Volume Conjecture for Knots*, SpringerBriefs in Mathematical Physics 30, https://doi.org/10.1007/978-981-13-1150-5_3

Conjecture 3.1 (Volume Conjecture) The following equality would hold for any knot K.

$$2\pi \lim_{N \to \infty} \frac{\log \left| J_N(K; \exp(2\pi \sqrt{-1}/N)) \right|}{N} = \mathrm{Vol}(S^3 \setminus K). \tag{3.1}$$

Here $\mathrm{Vol}(S^3 \setminus K)$ is the simplicial volume of the knot complement $S^3 \setminus K$ defined in Definition 1.6.

So far the volume conjecture is known to be true for the following knots and links.

- torus knots (Kashaev and Tirkkonen [40]),
- $(2, 2m)$ torus links (Hikami [32]),
- the figure-eight knot (Ekholm; See Sect. 3.2 for the proof),
- hyperbolic knot 5_2 (Kashaev and the second author [41], T. Ohtsuki [71]),
- hyperbolic knots 6_1, 6_2, and 6_3 (Ohtsuki and Yokota [72]),
- $(2, 2m + 1)$ cable of the figure-eight knot (Lê and A. Tran [52]),
- Whitehead doubles of torus knots (H. Zheng [96]),
- twisted Whitehead links (Zheng [96]),
- Borromean rings (S. Garoufalidis and Lê [25]),
- Whitehead chains (R. van der Veen [84]).

3.2 Figure-Eight Knot

In this section we follow Ekholm to give a proof of the volume conjecture for the figure-eight knot $\mathscr{E}$ (Figs. 1.19 and 1.16).

From (2.11), we have

$$\begin{aligned} J_N(\mathscr{E}; q) &= \frac{1}{\{N\}} \sum_{j=0}^{N-1} \frac{\{N+j\}!}{\{N-1-j\}!} \\ &= \sum_{j=0}^{N-1} \prod_{k=1}^{j} \left(q^{(N-k)/2} - q^{-(N-k)/2} \right) \left(q^{(N+k)/2} - q^{-(N+k)/2} \right). \end{aligned} \tag{3.2}$$

Replacing q with $\exp(2\pi \sqrt{-1}/N)$ we obtain

$$J_N(\mathscr{E}; \exp(2\pi \sqrt{-1}/N)) = \sum_{j=0}^{N-1} g_N(j), \tag{3.3}$$

where $g_N(j) = \prod_{k=1}^{j} 4 \sin^2(k\pi/N)$. Since a graph of $y = 4 \sin^2(\pi x)$ is as Fig. 3.1, we have the following:

Fig. 3.1 Graph of
$y = 4\sin^2(\pi x)$

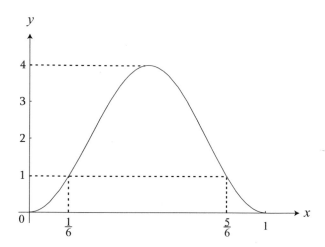

- When $0 < j < N/6$, $g_N(j)$ decreases,
- when $N/6 < j < 5N/6$, $g_N(j)$ increases, and
- when $5N/6 < j < N$, $g_N(j)$ decreases.

Therefore we see that $\max_{0 \le j < N}\{g_N(j)\}$ is attained at $j = \lfloor 5N/6 \rfloor$. Since $g_N(j) > 0$ we have

$$g_N(\lfloor 5N/6 \rfloor) < \sum_{j=0}^{N-1} g_N(j) < N g_N(\lfloor 5N/6 \rfloor).$$

Taking log and divide by N, we have

$$\frac{\log g_N(\lfloor 5N/6 \rfloor)}{N} < \frac{\log\left(\sum_{j=0}^{N-1} g_N(j)\right)}{N} < \frac{\log N}{N} + \frac{\log g_N(\lfloor 5N/6 \rfloor)}{N}.$$

Since $\lim_{N \to \infty} \frac{\log N}{N} = 0$, we have

$$\lim_{N \to \infty} \frac{\log\left(J_N(\mathscr{E}; \exp(2\pi\sqrt{-1}/N))\right)}{N} = \lim_{N \to \infty} \frac{\log g_N(\lfloor 5N/6 \rfloor)}{N}$$

$$= \lim_{N \to \infty} \sum_{j=1}^{\lfloor 5N/6 \rfloor} \frac{2\log\left(2\sin(j\pi/N)\right)}{N}$$

$$= \frac{2}{\pi} \int_0^{5\pi/6} \log(2\sin x)\,dx = -\frac{2}{\pi}\Lambda(5\pi/6),$$

where $\Lambda(\theta)$ is the Lobachevsky function (see (1.2)). The following formulas are well known (see for example [60]):

$$\Lambda(z+\pi) = \Lambda(z),$$
$$\Lambda(-z) = -\Lambda(z),$$
$$\Lambda(2z) = 2\Lambda(z) + 2\Lambda(z+\pi/2).$$

Therefore we have

$$\Lambda(5\pi/6) = \Lambda(\pi - \pi/6) = -\Lambda(\pi/6) = -\Lambda(\pi/3)/2 + \Lambda(2\pi/3) = -3\Lambda(\pi/3)/2.$$
$$(3.4)$$

So we finally have

$$\lim_{N\to\infty} \frac{\log\left|J_N(\mathscr{E}; e^{2\pi\sqrt{-1}/N})\right|}{N} = \frac{3}{\pi}\Lambda(\pi/3) = \frac{\mathrm{Vol}(S^3 \setminus \mathscr{E})}{2\pi},$$

proving the volume conjecture for the figure-eight knot.

3.3   Torus Knot

In this section we prove the volume conjecture for the torus knot $T(2, 2a + 1)$. To do this, we study the asymptotic behavior of the colored Jones polynomial $J_N\bigl(T(2, 2a + 1); e^{2\pi\sqrt{-1}/N}\bigr)$ for large N.

Let ξ be a complex variable near $2\pi\sqrt{-1}$. Multiplying by $q^{N/2} - q^{-N/2}$ and replacing q with $\exp(\xi/N)$ in (2.16), we have

$$(e^{\xi/2} - e^{-\xi/2})J_N(T(2, 2a + 1); e^{\xi/N})$$

$$= (-1)^{N-1} \exp\left(\frac{-(2a + 1)(N^2 - 1)\xi}{2N}\right)$$

$$\times \left(\sum_{c=0}^{N-1}(-1)^c \exp\left(\frac{((2a + 1)(c^2 + c) + 2c + 1)\xi}{2N}\right)\right.$$

$$\left. - \sum_{c=0}^{N-1}(-1)^c \exp\left(\frac{((2a + 1)(c^2 + c) - 2c - 1)\xi}{2N}\right)\right)$$

$$= (-1)^{N-1} \exp\left(\frac{-(2a + 1)(N^2 - 1)\xi}{2N}\right) \exp\left(-\frac{\xi}{4N}\left(\frac{2a + 1}{2} + \frac{2}{2a + 1}\right)\right)$$

$$(\Sigma_+(\xi) - \Sigma_-(\xi)),$$

where

$$\Sigma_\pm(\xi) := \sum_{c=0}^{N-1}(-1)^c \exp\left(\frac{(2a + 1)\xi}{2N}\left(c + \frac{1}{2} \pm \frac{1}{2a + 1}\right)^2\right).$$

Now we use the following formula:

$$\sqrt{\frac{\alpha}{\pi}} \int_{C_\theta} \exp(-\alpha x^2 + px)\,dx = \exp\left(\frac{p^2}{4\alpha}\right),$$

where C_θ is the line $\{t\exp(\theta\sqrt{-1}) \mid t \in \mathbb{R}\}$. We choose θ so that $\mathrm{Re}(\alpha\exp(2\theta\sqrt{-1})) > 0$ to make the integral converge.

Putting $\alpha := \frac{N}{2(2a+1)\xi}$, $p := c + \frac{1}{2} \pm \frac{1}{2a+1}$, and $\theta := \pi/4$, we have

$$\Sigma_\pm(\xi)$$

$$= \sqrt{\frac{2N}{2(2a+1)\xi\pi}} \sum_{c=0}^{N-1} (-1)^c \int_{C_{\pi/4}} \exp\left(\frac{-N}{2(2a+1)\xi}x^2 + \left(c+\frac{1}{2}\pm\frac{1}{2a+1}\right)x\right)dx$$

$$= \sqrt{\frac{N}{2(2a+1)\xi\pi}}$$

$$\int_{C_{\pi/4}} \exp\left(\frac{-N}{2(2a+1)\xi}x^2 + \frac{x}{2} \pm \frac{x}{2a+1}\right)\left(\sum_{c=0}^{N-1}(-1)^c\exp(cx)\right)dx$$

$$= \sqrt{\frac{N}{2(2a+1)\xi\pi}} \int_{C_{\pi/4}} \exp\left(\frac{-N}{2(2a+1)\xi}x^2\right)\exp\left(\frac{\pm x}{2a+1}\right)\left(\frac{1-(-1)^N e^{Nx}}{e^{x/2}+e^{-x/2}}\right)dx.$$

Therefore we have

$$\Sigma_+(\xi) - \Sigma_-(\xi)$$

$$= \sqrt{\frac{N}{2(2a+1)\xi\pi}}$$

$$\times \left(\int_{C_{\pi/4}} \frac{\sinh\left(\frac{x}{2a+1}\right)}{\cosh\left(\frac{x}{2}\right)} \exp\left[\frac{-N}{2(2a+1)\xi}x^2\right]dx\right.$$

$$\left. - (-1)^N \int_{C_{\pi/4}} \frac{\sinh\left(\frac{x}{2a+1}\right)}{\cosh\left(\frac{x}{2}\right)} \exp\left[\frac{-N}{2(2a+1)\xi}x\right]\exp(Nx)\,dx\right)$$

$$= -(-1)^N \sqrt{\frac{N}{2(2a+1)\xi\pi}} \int_{C_{\pi/4}} \frac{\sinh\left(\frac{x}{2a+1}\right)}{\cosh\left(\frac{x}{2}\right)} \exp\left(\frac{-N}{2(2a+1)\xi}x^2 + Nx\right)dx,$$

since the integrand of the first integral is an odd function. So we have

$$(e^{\xi/2} - e^{-\xi/2}) J_N(T(2, 2a + 1); e^{\xi/N})$$

$$= \exp\left(\frac{-(2a+1)(N^2-1)\xi}{2N}\right) \exp\left(-\frac{\xi}{4N}\left(\frac{2a+1}{2} + \frac{2}{2a+1}\right)\right) \sqrt{\frac{N}{2(2a+1)\xi\pi}}$$

$$\times \int_{C_{\pi/4}} \frac{\sinh\left(\frac{x}{2a+1}\right)}{\cosh\left(\frac{x}{2}\right)} \exp\left(\frac{-N}{2(2a+1)\xi}x^2 + Nx\right) dx.$$

$$(3.5)$$

Taking the derivative with respect to ξ at $2\pi\sqrt{-1}$, we have

$$J_N(T(2, 2a + 1); e^{2\pi\sqrt{-1}/N})$$

$$= (-1)^{N+1} \exp\left(\frac{\pi\sqrt{-1}}{2N}\left(2(2a+1) - \frac{2a+1}{2} - \frac{2}{2a+1}\right)\right) \sqrt{\frac{N}{4(2a+1)\pi^2\sqrt{-1}}}$$

$$\times \int_{C_{\pi/4}} \frac{\sinh\left(\frac{x}{2a+1}\right)}{\cosh\left(\frac{x}{2}\right)} \frac{N}{2(2a+1)(2\pi\sqrt{-1})^2} x^2$$

$$\exp\left(\frac{-N}{2(2a+1)2\pi\sqrt{-1}}x^2 + Nx\right) dx$$

$$= (-1)^N \exp\left(\frac{\pi\sqrt{-1}}{2N}\left(2(2a+1) - \frac{2a+1}{2} - \frac{2}{2a+1}\right)\right)$$

$$\frac{N^{3/2}}{16(2a+1)^{3/2}\pi^3 e^{\pi\sqrt{-1}/4}}$$

$$\times \int_{C_{\pi/4}} \frac{\sinh\left(\frac{x}{2a+1}\right)}{\cosh\left(\frac{x}{2}\right)} x^2 \exp\left(\frac{-N}{4(2a+1)\pi\sqrt{-1}}x^2 + Nx\right) dx,$$

$$(3.6)$$

since the derivative at $\xi = 2\pi\sqrt{-1}$ of the integral in the right hand side of (3.5) should vanish.

Now we will use a special case of the *saddle point method* (see for example [55, Theorems 7.2.9]).

Theorem 3.1 *For a non-zero complex number a and a real number θ with* $\mathrm{Re}(a^{-1}\exp(2\theta\sqrt{-1})) > 0$, *we have*

$$\int_{C_\theta} g(x)e^{-Nx^2/a} dx = \sqrt{\frac{a\pi}{N}} g(0) + O(N^{-1}),$$

when $N \to \infty$.

Note that the assumption $\mathrm{Re}(a^{-1}\exp(2\theta\sqrt{-1})) > 0$ is to make the integral converge.

Using Theorem 3.1 we will calculate the integral in (3.6). We have

$$
\int_{C_{\pi/4}} \frac{x^2 \sinh\left(\frac{x}{2a+1}\right)}{\cosh\left(\frac{x}{2}\right)} \exp\left(\frac{-N}{4(2a+1)\pi\sqrt{-1}}x^2 + Nx\right) dx
$$

$$
=(-1)^N \int_{C_{\pi/4}} \frac{x^2 \sinh\left(\frac{x}{2a+1}\right)}{\cosh\left(\frac{x}{2}\right)} \exp\left(\frac{-N}{4(2a+1)\pi\sqrt{-1}}(x-2(2a+1)\pi\sqrt{-1})^2\right) dx.
$$

Let $\tilde{C}_{\pi/4}$ be the line $\{t \exp(\pi\sqrt{-1}/4) + 2(2a+1)\pi\sqrt{-1} \mid t \in \mathbb{R}\}$. Then by the residue theorem we have

$$
\int_{C_{\pi/4}} \frac{x^2 \sinh\left(\frac{x}{2a+1}\right)}{\cosh\left(\frac{x}{2}\right)} \exp\left(\frac{-N}{4(2a+1)\pi\sqrt{-1}}(x-2(2a+1)\pi\sqrt{-1})^2\right) dx.
$$

$$
=\int_{\tilde{C}_{\pi/4}} \frac{x^2 \sinh\left(\frac{x}{2a+1}\right)}{\cosh\left(\frac{x}{2}\right)} \exp\left(\frac{-N}{4(2a+1)2\pi\sqrt{-1}}(x-2(2a+1)\pi\sqrt{-1})^2\right) dx
$$

$$
+2\pi\sqrt{-1}
$$

$$
\times \sum_k \mathrm{Res}\left(\frac{x^2 \sinh\left(\frac{x}{2a+1}\right)}{\cosh\left(\frac{x}{2}\right)} \exp\left(\frac{-N}{4(2a+1)\pi\sqrt{-1}}(x-2(2a+1)\pi\sqrt{-1})^2\right);\right.
$$

$$
\left. x = (2k+1)\pi\sqrt{-1}\right)
$$

$$
=\int_{\tilde{C}_{\pi/4}} \frac{x^2 \sinh\left(\frac{x}{2a+1}\right)}{\cosh\left(\frac{x}{2}\right)} \exp\left(\frac{-N}{4(2a+1)\pi\sqrt{-1}}(x-2(2a+1)\pi\sqrt{-1})^2\right) dx
$$

$$
+2\pi\sqrt{-1}\sum_{k=0}^{2a}(-1)^{k+1}2\sqrt{-1}((2k+1)\pi\sqrt{-1})^2 \sinh\left(\frac{(2k+1)\pi\sqrt{-1}}{2a+1}\right)
$$

$$
\times \exp\left(\frac{-N}{4(2a+1)\pi\sqrt{-1}}((2k+1)\pi\sqrt{-1}-2(2a+1)\pi\sqrt{-1})^2\right)
$$

$$
=\int_{\tilde{C}_{\pi/4}} \frac{x^2 \sinh\left(\frac{x}{2a+1}\right)}{\cosh\left(\frac{x}{2}\right)} \exp\left(\frac{-N}{4(2a+1)\pi\sqrt{-1}}(x-2(2a+1)\pi\sqrt{-1})^2\right) dx
$$

$$
+4\pi^3 \sum_{k=0}^{2a}(-1)^{k+1}(2k+1)^2 \sinh\left(\frac{(2k+1)\pi\sqrt{-1}}{2a+1}\right) \exp\left(\frac{(4a-2k+1)^2\pi N}{4(2a+1)\sqrt{-1}}\right),
$$

where $\mathrm{Res}(f(x); x = x_0)$ is the residue of $f(x)$ at x_0.

Putting $y := x - 2(2a + 1)\pi\sqrt{-1}$ the integral becomes

$$-\int_{C_{\pi/4}} \frac{(y + 2(2a + 1)\pi\sqrt{-1})^2 \sinh\left(\frac{y}{2a+1}\right)}{\cosh\left(\frac{y}{2}\right)}$$

$$\exp\left(\frac{-N}{4(2a + 1)\pi\sqrt{-1}}y^2\right) dy = O(N^{-1})$$

from Theorem 3.1. Therefore we finally have the following equality.

$$J_N\left(T(2, 2a + 1); e^{2\pi\sqrt{-1}/N}\right)$$

$$= \exp\left(\frac{\pi\sqrt{-1}}{2N}\left(2(2a + 1) - \frac{2a + 1}{2} - \frac{2}{2a + 1}\right)\right) \frac{N^{3/2}}{4(2a + 1)^{3/2}e^{\pi\sqrt{-1}/4}}$$

$$\times \sum_{k=0}^{2a}(-1)^{k+1}(2k + 1)^2 \sinh\left(\frac{(2k + 1)\pi\sqrt{-1}}{2a + 1}\right)\exp\left(\frac{(4a - 2k + 1)^2\pi N}{4(2a + 1)\sqrt{-1}}\right)$$

$$+ O(N^{1/2}).$$

$$(3.7)$$

This means that $J_N\left(T(2, 2a + 1); e^{2\pi\sqrt{-1}/N}\right)$ grows polynomially and so we have

$$\lim_{N\to\infty} \frac{\log\left|J_N\left(T(2, 2a + 1); e^{2\pi\sqrt{-1}/N}\right)\right|}{N} = 0.$$

Since it is known that the complement of any torus knot is Seifert fibered, the volume conjecture for $T(2, 2a + 1)$ follows.

Chapter 4
Idea of "Proof"

Abstract In this chapter, for a hyperbolic knot K, we explain an idea of a possible proof of the Volume Conjecture by using Kashaev's invariant $\langle K \rangle_N$ of K, which is known to be the N-colored Jones polynomial $J_N(K, q)$ evaluated at

$$q = \exp \frac{2\pi \sqrt{-1}}{N}$$

after the work of Murakami and Murakami (Acta Math 186(1):85–104, 2001. MR 1828373). By using $\langle K \rangle_N$ rather than $J_N(K, q)$, we can observe the correspondence between the algebraic structure of $\langle K \rangle_N$ and the geometric structure of the complement of K more clearly. Throughout this chapter, we set q as above. In the first section, following Yokota (Interdiscip Inf Sci 9(1):11–21, 2003. MR MR2023102 (2004j:57014)), we explain how to compute the invariant and how to reduce it. In the second section, following Kashaev and Yokota (On the volume conjecture for 5_2, Preprint, 2012), we explain how to compute the asymptotic behavior of an integral expression of the invariant. In the third section, following Yokota (Interdiscip Inf Sci 9(1):11–21, 2003. MR MR2023102 (2004j:57014)) again, we explain the relationship between the hyperbolic structure of the knot complement, and a "potential" function which we obtain in the second section. In the fourth section, we sort the remaining tasks.

4.1 Algebraic Part

For simplicity, we put $\mathcal{N} = \{0, 1, \ldots, N - 1\}$ and

$$\theta_{kl}^{ij} = \begin{cases} 1 & \text{if } [i - j] + [j - l] + [l - k - 1] + [k - i] = N - 1, \\ 0 & \text{otherwise} \end{cases}$$

for $i, j, k, l \in \mathcal{N}$, where $[v] \in \mathcal{N}$ denotes the residue of v modulo N. Note that

$$[i - j] + [j - l] + [l - k - 1] + [k - i] \equiv -1 \mod N$$

H. Murakami, Y. Yokota, *Volume Conjecture for Knots*, SpringerBriefs in Mathematical Physics 30, https://doi.org/10.1007/978-981-13-1150-5_4

and that both $[i - j - 1] + [k - l]$ and $[i - l] + [k - j]$ are less than N if and only if $\theta_{kl}^{ij} = 1$, that is,

$$\theta_{kl}^{ij} = \theta([i - j] + [k - l - 1]) \cdot \theta([j - l] + [k - i]).$$

Thus, if we define the q-factorials $(q)_\nu$ and $(\bar{q})_\nu$ by

$$(q)_\nu = (1 - q)(1 - q^2) \cdots (1 - q^{[\nu]}), \quad (\bar{q})_\nu = (1 - \bar{q})(1 - \bar{q}^2) \cdots (1 - \bar{q}^{[\nu]}),$$

the R-matrices in Sect. 2.1.3 can be rewritten as

$$(R_\mathrm{K})_{kl}^{ij} = \frac{Nq^{-\frac{1}{2}+i-k}\theta_{kl}^{ij}}{(q)_{i-j}(\bar{q})_{j-l}(q)_{l-k-1}(\bar{q})_{k-i}}, \quad \left(R_\mathrm{K}^{-1}\right)_{kl}^{ij} = \frac{Nq^{\frac{1}{2}+j-l}\theta_{kl}^{ij}}{(\bar{q})_{i-j}(q)_{j-l}(\bar{q})_{l-k-1}(q)_{k-i}},$$

where we used $(q)_\nu = (-1)^\nu q^{\nu(\nu+1)/2}(\bar{q})_\nu$. Then, Kashaev's invariant $\langle K \rangle_N$ of K is obtained by contracting the tensors

$$(R_\mathrm{K})_{kl}^{ij}, \quad \left(R_\mathrm{K}^{-1}\right)_{kl}^{ij}, \quad -q^{-\frac{1}{2}}\delta_{k+1,l}, \quad q^{\frac{1}{2}}\delta_{i-1,j}$$

associated to the critical points, which are depicted in Fig. 4.1, of the $(1,1)$-tangle presentation of K respectively.

Example 4.1 Let K denote the knot 6_1 depicted in Fig. 4.2, where the broken edge is labeled 0. Then, we only consider the labelling satisfying

$$\theta_{an}^{(i+1)0} \cdot \theta_{ij}^{ac} \cdot \theta_{cd}^{nm} \cdot \theta_{l(m+1)}^{(b+1)d} \cdot \theta_{be}^{lk} \cdot \theta_{0(k+1)}^{je} \neq 0,$$

that is,

$$6(N - 1) = ([i + 1] + [a - i - 1] + [n - a - 1] + [-n])$$
$$+ ([a - c] + [i - a] + [j - i - 1] + [c - j])$$
$$+ ([n - m] + [c - n] + [d - c - 1] + [m - d])$$
$$+ ([b + 1 - d] + [l - b - 1] + [m - l] + [d - m - 1])$$

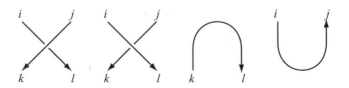

Fig. 4.1 Critical points of a $(1,1)$-tangle presentation of K

Fig. 4.2 A (1,1)-tangle
presentation of 6_1

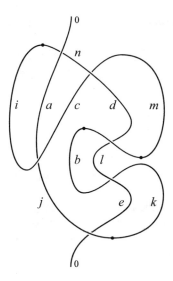

$$+ ([l - k] + [b - l] + [e - b - 1] + [k - e])$$
$$+ ([j - e] + [-j] + [k] + [e - k - 1]).$$

The right hand side can be rewritten as

$$[i + 1] + [j - i - 1] + [-j]$$
$$+ [a - l - 1] + [i - a]$$
$$+ [n - a - 1] + [c - n] + [a - c]$$
$$+ [d - c - 1] + [c - j] + [j - e] + [e - b - 1] + [b + 1 - d]$$
$$+ [m - d] + [d - m - 1]$$
$$+ [l - b - 1] + [b - l]$$
$$+ [k - e] + [e - k - 1]$$
$$+ [k] + [l - k] + [m - l] + [n - m] + [-n]$$
$$\geq [i + 1] + [j - i - 1] + [-j] + 6(N - 1)$$
$$+ [k] + [l - k] + [m - l] + [n - m] + [-n],$$

where each line in the left hand side corresponds to each face of Fig. 4.2, and so we
have

$$[i + 1] + [j - i - 1] + [-j] = 0 = [k] + [l - k] + [m - l] + [n - m] + [-n],$$

that is, $i + 1 = j = k = l = m = n = 0$, and

$$
\begin{aligned}
N - 1 &= [a] + [-1 - a] \\
&= [-a - 1] + [c] + [a - c] \\
&= [d - c - 1] + [c] + [-e] + [e - b - 1] + [b + 1 - d] \\
&= [-d] + [d - 1] \\
&= [-b - 1] + [b] \\
&= [-e] + [e - 1],
\end{aligned}
$$

which implies $0 \le c \le a < N$ and $c < d \le b + 1 \le e \le N$. Therefore, we can ignore the q-factorials corresponding to the corners in the unbounded regions in Fig. 4.2, and $\langle K \rangle_N$ is given by

$$
\langle K \rangle_N = \sum_{c \le d' \le b} \frac{Nq^{\frac{1}{2} + d'}}{(\bar{q})_{b - d'} (q)_{d'} (q)_{N-1-b}} \cdot \frac{Nq^{\frac{1}{2} - d' - 1}}{(q)_{N-1-d'} (\bar{q})_{d' - c} (q)_c}
$$

$$
\times \left(\sum_{a=c}^{N-1} \frac{Nq^{-\frac{1}{2} - a}}{(q)_{N-1-a} (\bar{q})_a} \cdot \frac{Nq^{-\frac{1}{2} + a + 1}}{(q)_{a-c} (\bar{q})_c (\bar{q})_{N-1-a}} \right)
$$

$$
\times \left(\sum_{e'=b}^{N-1} \frac{Nq^{\frac{1}{2} + e'}}{(q)_{e'} (\bar{q})_{N-1-e'}} \cdot \frac{Nq^{\frac{1}{2} - e' - 1}}{(q)_{N-1-e'} (\bar{q})_{e' - b} (q)_b} \right),
$$

where we put $d' = d - 1$ and $e' = e - 1$. Then, we apply the following lemma due to [66].

Lemma 4.1 *For any $\nu \in \mathcal{N}$, we have $(q)_\nu (\bar{q})_{N-1-\nu} = N$. Furthermore,*

$$
\sum_{\alpha \le \nu \le \beta} \frac{1}{(q)_{\beta - \nu} (\bar{q})_{\nu - \alpha}} = 1,
$$

where $0 \le \alpha \le \beta < N$.

By using Lemma 4.1, we can further eliminate the q-factorials around the edges labeled a and e in Fig. 4.2, and $\langle K \rangle_N$ is equal to

$$
\sum_{c \le d' \le b} \frac{Nq^{\frac{1}{2} + d'}}{(\bar{q})_{b - d'} (q)_{d'} (q)_{N-1-b}} \cdot \frac{Nq^{\frac{1}{2} - d' - 1}}{(q)_{N-1-d'} (\bar{q})_{d' - c} (q)_c} \cdot \frac{N^2}{(\bar{q})_c (q)_b}
$$

$$
= N^4 \sum_{c \le d' \le b} \frac{1}{(\bar{q})_{b - d'} (q)_{d'} (q)_{N-1-b} (q)_{N-1-d'} (\bar{q})_{d' - c} (q)_c (\bar{q})_c (q)_b}.
$$

Fig. 4.3 The twist knot with
$n + 3$ crossings

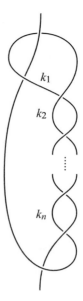

Similarly, Kashaev's invariant of the twist knot with $n + 3$ crossings, which is depicted in Fig. 4.3, is given by

$$
N^{n+1} \sum_{0 \le k_1 \le \cdots \le k_n < N} \frac{1}{(\bar{q})_{k_1} (q)_{k_n}} \prod_{\nu=1}^{n-1} \frac{1}{(q)_{k_\nu} (\bar{q})_{k_{\nu+1} - k_\nu} (q)_{N-1-k_{\nu+1}}}.
$$

As Example 4.1 suggests, in general, Kashaev's invariant of a knot K in S^3 can be written as a sum of products of q-factorials each of which corresponds to a corner, which is not in the unbounded regions or not around the edges located at the "entrance" and the "exit", of a $(1,1)$-tangle presentation of K.

4.2 Analytic Part

In this section, we explain how to compute the asymptotic behavior of Kashaev's invariant. In the first subsection, we replace the q-factorials in the invariant with quantum dilogarithm functions due to L. Faddeev [23], and give an integral expression of the invariant by using the residue theorem. In the second subsection, we explain the asymptotic behavior of quantum dilogarithm functions, and derive a "potential" function from the invariant. In the third subsection, we explain how to apply the saddle point method to the integral expression of the invariant.

4.2.1   *Integral Expression*

We define Faddeev's *quantum dilogarithm* function by

$$\psi_N(z) = \exp \frac{1}{4} \int_{-\infty}^{\infty} \frac{e^{(2z-1)t} \, dt}{t \sinh t \cdot \sinh t N^{-1}}$$

in the strip $|\operatorname{Re}(2z-1)| < 1 + N^{-1}$, where the singularity at $t = 0$ is put below the contour of integration, which can be extended to a meromorphic function on $\mathbb{C}$ by the functional equation

$$\frac{\psi_N(z+1)}{\psi_N(z)} = \frac{1}{1 + e^{2\pi\sqrt{-1}Nz}}.$$

Note that the sets of poles of ψ_N is given by $\{p_k : k \geq N\}$ and the set of zeroes of ψ_N is $\{p_k : k < 0\}$, where we put

$$p_k = \frac{2k+1}{2N}.$$

In the first place, in [23], ψ_N is given as a solution to the functional equation

$$\frac{\psi_N(z+p_0)}{\psi_N(z-p_0)} = \frac{1}{1 - e^{2\pi\sqrt{-1}z}}, \tag{4.1}$$

which implies the important identities

$$\frac{1}{(q)_k} = \frac{\psi_N(p_1)}{\psi_N(p_0)} \cdots \frac{\psi_N(p_k)}{\psi_N(p_{k-1})} = \frac{\psi_N(p_k)}{\psi_N(p_0)},$$

$$\frac{1}{(\bar{q})_k} = \frac{\psi_N(1-p_0)}{\psi_N(1-p_1)} \cdots \frac{\psi_N(1-p_{k-1})}{\psi_N(1-p_k)} = \frac{\psi_N(1-p_0)}{\psi_N(1-p_k)}$$

for our purpose. In fact, the left hand side is equal to

$$\exp \frac{1}{2} \int_{-\infty}^{\infty} \frac{e^{(2z-1)t}}{t \sinh t} dt = \exp \frac{1}{2} \left\{ 2\pi\sqrt{-1} \sum_{k=1}^{\infty} \operatorname*{Res}_{t=k\pi\sqrt{-1}} \frac{e^{(2z-1)t}}{t \sinh t} \right\}$$

$$= \exp \left\{ \sum_{k=1}^{\infty} \frac{(e^{2\pi\sqrt{-1}z})^k}{k} \right\} = \exp \left\{ \log(1 - e^{2\pi\sqrt{-1}z}) \right\},$$

where we used

$$\operatorname*{Res}_{t=k\pi\sqrt{-1}} \frac{e^{(2z-1)t}}{t \sinh t} = \lim_{t \to k\pi\sqrt{-1}} \frac{e^{(2z-1)t}(t - k\pi\sqrt{-1})}{t \sinh t} = \frac{(e^{2\pi\sqrt{-1}z})^k}{k\pi\sqrt{-1}}.$$

Note also that $\psi_N(p_0)$ and $\psi_N(1 - p_0)$ are explicitly given by

$$\psi_N(p_0) = \sqrt{N} \exp \frac{N}{2\pi\sqrt{-1}} \left(\frac{\pi^2}{6} - \frac{\pi^2}{2N} + \frac{\pi^2}{6N^2} \right),$$

$$\psi_N(1 - p_0) = \frac{1}{\sqrt{N}} \exp \frac{N}{2\pi\sqrt{-1}} \left(\frac{\pi^2}{6} - \frac{\pi^2}{2N} + \frac{\pi^2}{6N^2} \right). \tag{4.2}$$

Then, we can rewrite Kashaev's invariant by using quantum dilogarithm functions.

Example 4.2 If K is the twist knot with $n + 3$ crossings, $\langle K \rangle_N$ is equal to

$$N^{n+1} \sum_{k_1 \le \cdots \le k_n} \frac{\psi_N(1 - p_0)}{\psi_N(1 - p_{k_1})} \frac{\psi_N(p_k)}{\psi_N(p_0)} \prod_{\nu=1}^{n-1} \frac{\psi_N(p_{k_\nu})}{\psi_N(p_0)} \frac{\psi_N(1 - p_0)}{\psi_N(1 - p_{k_{\nu+1}-k_\nu})} \frac{\psi_N(1 - p_{k_{\nu+1}})}{\psi_N(p_0)}.$$

Since $\psi_N(1 - p_{k_{\nu+1}-k_\nu}) = \infty$ if $k_{\nu+1} < k_\nu$, we can write

$$\langle K_n \rangle_N = N^{\frac{3-n}{2}} \sum_{k_1=0}^{N-1} \cdots \sum_{k_n=0}^{N-1} \Psi_N(p_{k_1}, \ldots, p_{k_n}),$$

where $\Psi_N(z_1, \ldots, z_n)$ is given by

$$e^{\frac{N(n-1)}{2\pi\sqrt{-1}}\left(-\frac{\pi^2}{6} + \frac{\pi^2}{2N} - \frac{\pi^2}{6N^2}\right)} \frac{\psi_N(z_n)}{\psi_N(1 - z_1)} \prod_{\nu=1}^{n-1} \frac{\psi_N(z_\nu)\psi_N(1 - z_{\nu+1})}{\psi_N(1 - z_{\nu+1} + z_\nu - p_0)}.$$

As Example 4.2 suggests, in general, Kashaev's invariant of a knot K in S^3 can be written as

$$\langle K \rangle_N = N^{\frac{3-n}{2}} \sum_{k_1=0}^{N-1} \cdots \sum_{k_n=0}^{N-1} \Psi_N(p_{k_1}, \ldots, p_{k_n}),$$

where $\Psi_N(z_1, \ldots, z_n)$ is a product of quantum dilogarithm functions each of which corresponds to a corner, which is not in the unbounded regions and not around the edges located at the "entrance" and the "exit", of a $(1,1)$-tangle presentation of K. Then, by the residue theorem, we have

$$\langle K \rangle_N = (-1)^n N^{\frac{n+3}{2}} \int_C \frac{dz_1}{1 + Q_1^N} \cdots \int_C \frac{dz_n}{1 + Q_n^N} \Psi_N(z_1, \ldots, z_n),$$

where we put $Q_\nu = e^{2\pi\sqrt{-1}z_\nu}$ and $C = \{z \in \mathbb{C} : |z - \frac{1}{2}| = \frac{1}{2}\}$. Furthermore, if we decompose C into

$$A = \{z \in C : \operatorname{Im} z \ge 0\}, \qquad B = \{z \in C : \operatorname{Im} z \le 0\},$$

we can observe

$$\int_C \frac{dz_\nu}{1 + Q_\nu^N} = \int_A \frac{dz_\nu}{1 + Q_\nu^N} + \int_B \frac{dz_\nu}{1 + Q_\nu^N}$$

$$= \int_1^0 dz_\nu - \int_A \frac{Q_\nu^N}{1 + Q_\nu^N} dz_\nu + \int_B \frac{Q_\nu^{-N}}{1 + Q_\nu^{-N}} dz_\nu$$

$$= \int_1^0 (1 - Q_\nu^N - Q_\nu^{-N}) dz_\nu + \int_A \frac{Q_\nu^{2N}}{1 + Q_\nu^N} dz_\nu - \int_B \frac{Q_\nu^{-2N}}{1 + Q_\nu^{-N}} dz_\nu = \cdots$$

and, roughly speaking, we can write

$$\langle K \rangle_N = (-1)^n \sum_{\varepsilon_1, \dots, \varepsilon_n} \langle K; \varepsilon_1, \dots, \varepsilon_n \rangle_N,$$

where we put

$$\langle K; \varepsilon_1, \dots, \varepsilon_n \rangle_N = N^{\frac{n+3}{2}} \int_0^1 dz_1 \cdots \int_0^1 dz_n \, \Psi_N(z_1, \dots, z_n) \prod_{\nu=1}^n (-Q_\nu^N)^{\varepsilon_\nu}.$$

In practice, for our purpose, a finite expansion of $\langle K \rangle_N$ is sufficient.

4.2.2 Potential Function

The asymptotic behavior of ψ_N is described by Euler's dilogarithm

$$\mathrm{Li}_2(z) = - \int_0^z \frac{\log(1 - t)}{t} dt.$$

In fact, we can observe that

$$\psi_N(z) = \exp \frac{1}{4} \int_{-\infty}^\infty \frac{N e^{(2z-1)t}}{t^2 \sinh t} \cdot \frac{t N^{-1}}{\sinh t N^{-1}} dt$$

$$= \exp \frac{1}{4} \left\{ \int_{-\infty}^\infty \frac{N e^{(2z-1)t}}{t^2 \sinh t} dt + O(N^{-1}) \right\}$$

$$= \exp \frac{1}{4} \left\{ 2\pi \sqrt{-1} \sum_{k=1}^\infty \operatorname*{Res}_{t = k\pi\sqrt{-1}} \frac{N e^{(2z-1)t}}{t^2 \sinh t} + O(N^{-1}) \right\},$$

where

$$\operatorname*{Res}_{t = k\pi\sqrt{-1}} \frac{e^{(2z-1)t}}{t^2 \sinh t} = \lim_{t \to k\pi\sqrt{-1}} \frac{e^{(2z-1)t}(t - k\pi\sqrt{-1})}{t^2 \sinh t} = \frac{(e^{2\pi\sqrt{-1}z})^k}{-k^2\pi^2},$$

and that

$$\psi_N(z) = \exp\left\{\frac{N}{2\pi\sqrt{-1}}\sum_{k=1}^{\infty}\frac{(e^{2\pi\sqrt{-1}z})^k}{k^2} + O(N^{-1})\right\}$$

$$= \exp\left\{\frac{N}{2\pi\sqrt{-1}}\operatorname{Li}_2(e^{2\pi\sqrt{-1}z}) + O(N^{-1})\right\}$$

if $\operatorname{Im} z \geq 0$. Similarly, if $\operatorname{Im} z < 0$, we have

$$\psi_N(z) = \exp\frac{1}{4}\left\{-2\pi\sqrt{-1}\sum_{k=0}^{\infty}\operatorname*{Res}_{t=-k\pi\sqrt{-1}}\frac{Ne^{(2z-1)t}}{t^2\sinh t} + O(N^{-1})\right\}$$

$$= \exp\left\{\frac{N}{2\pi\sqrt{-1}}\left(\operatorname*{Res}_{t=0}\frac{Ne^{(2z-1)t}}{t^2\sinh t} - \sum_{k=1}^{\infty}\frac{(e^{2\pi\sqrt{-1}z})^{-k}}{k^2}\right) + O(N^{-1})\right\}$$

$$= \exp\left\{\frac{N}{2\pi\sqrt{-1}}\left(2\pi^2 z(z-1) + \frac{\pi^2}{3} - \operatorname{Li}_2(e^{-2\pi\sqrt{-1}z})\right) + O(N^{-1})\right\}.$$

Consequently, by the functional equation above, we can observe

$$\psi_N(z) = \exp\left\{\frac{N\mathscr{L}(z)}{2\pi\sqrt{-1}} + O(N^{-1})\right\} \tag{4.3}$$

for any $z \in \mathbb{C} - \{(-\infty, 0) \cup (1, \infty)\}$, where

$$\mathscr{L}(z) = \begin{cases} \operatorname{Li}_2(e^{2\pi\sqrt{-1}z}) & \text{if } \operatorname{Im} z \geq 0, \\ -\operatorname{Li}_2(e^{-2\pi\sqrt{-1}z}) + 2\pi^2 z(z-1) + \frac{1}{3}\pi^2 & \text{if } \operatorname{Im} z < 0. \end{cases}$$

Note that $\mathscr{L}(z)$ is analytic in $\mathbb{C} - \{(-\infty, 0] \cup [1, \infty)\}$.

Example 4.3 Suppose K is the twist knot with $n + 3$ crossings. Since

$$\psi_N(z \pm p_0) = \exp\left\{\frac{N\mathscr{L}(z)}{2\pi\sqrt{-1}} \mp \frac{1}{2}\log(1 - e^{2\pi\sqrt{-1}z}) + O(N^{-1})\right\},$$

we have

$$\Psi_N(z_1, \ldots, z_n) =$$

$$\exp\left\{\frac{NH(z_1, \ldots, z_n)}{2\pi\sqrt{-1}} - \frac{1}{2}\sum_{v=1}^{n-1}\log(1 - e^{2\pi\sqrt{-1}(1-z_{v+1}+z_v)}) + O(N^{-1})\right\},$$

where $H(z_1, \ldots, z_n)$ is defined by

$$\mathscr{L}(z_n) - \mathscr{L}(1 - z_1) + \sum_{\nu=1}^{n-1} \{\mathscr{L}(z_\nu) - \mathscr{L}(1 - z_{\nu+1} + z_\nu) + \mathscr{L}(1 - z_{\nu+1})\} - \frac{1}{6}(n-1)\pi^2.$$

As Example 4.3 suggests, in general, Ψ_N which appears in the integral formula of Kashaev's invariant can be written as

$$\Psi_N(z_1, \ldots, z_n) = \exp\left\{\frac{NH(z_1, \ldots, z_n)}{2\pi\sqrt{-1}} + T(z_1, \ldots, z_n) + O(N^{-1})\right\},$$

where $T(z_1, \ldots, z_n)$ is a sum of logarithms and $H(z_1, \ldots, z_n)$, called *potential function* in what follows, is a sum of Euler's dilogarithms each of which corresponds to a corner, which is not in the unbounded regions and not around the edges located at the "entrance" and the "exit", of a $(1,1)$-tangle presentation of K.

4.2.3   Saddle Point Method

To estimate $\langle K; \varepsilon_1, \ldots, \varepsilon_n \rangle_N$, we are going to find a domain $\Omega_{\varepsilon_1, \ldots, \varepsilon_n} \subset [0,1]^n$ such that, under some assumptions,

$$\langle K \rangle_N^{\Omega_{\varepsilon_1, \ldots, \varepsilon_n}} = N^{\frac{n+3}{2}} \int_{\Omega_{\varepsilon_1, \ldots, \varepsilon_n}} \Psi_N(z_1, \ldots, z_n) \prod_{\nu=1}^{n} (-Q_\nu^N)^{\varepsilon_\nu} dz_1 \cdots dz_n$$

is asymptotically equal to

$$N^{\frac{3}{2}} \exp \frac{NH_{\varepsilon_1, \ldots, \varepsilon_n}(\zeta_{\varepsilon_1, \ldots, \varepsilon_n})}{2\pi\sqrt{-1}},$$

where $\zeta_{\varepsilon_1, \ldots, \varepsilon_n}$ denotes a critical point of a branch of the potential function

$$H_{\varepsilon_1, \ldots, \varepsilon_n}(z_1, \ldots, z_n) = H(z_1, \ldots, z_n) - 4\pi^2(\varepsilon_1 z_1 + \cdots + \varepsilon_n z_n)$$

whose imaginary part is denoted by $f_{\varepsilon_1, \ldots, \varepsilon_n}(z_1, \ldots, z_n)$ for simplicity.

A candidate of $\Omega_{\varepsilon_1, \ldots, \varepsilon_n}$ is the set $\Delta_{\varepsilon_1, \ldots, \varepsilon_n}$ of $(\xi_1, \ldots, \xi_n) \in [0,1]^n$ such that

$$\lim_{\eta_1^2 + \cdots + \eta_n^2 \to \infty} f_{\varepsilon_1, \ldots, \varepsilon_n}(\xi_1 + \eta_1\sqrt{-1}, \ldots, \xi_n + \eta_n\sqrt{-1}) = \infty$$

and that $f_{\varepsilon_1, \ldots, \varepsilon_n}(z_1, \ldots, z_n)$ is smooth at $(z_1, \ldots, z_n) = (\xi_1, \ldots, \xi_n)$. Note that we can compute $\Delta_{\varepsilon_1, \ldots, \varepsilon_n}$ explicitly because

$$\operatorname{Im} \mathscr{L}(z) \sim \alpha(y_z)(x_z - \pi)$$

if $|y_z| \to \infty$, where we put $x_z = \operatorname{Re} 2\pi z$, $y_z = \operatorname{Im} 2\pi z$ and define $\alpha : \mathbb{R} \to \mathbb{R}$ by

$$\alpha(y) = \frac{1}{2}(y - |y|).$$

Example 4.4 Suppose K is the twist knot with 6 crossings. Then,

$$
\begin{aligned}
f_{\varepsilon_1,\varepsilon_2,\varepsilon_3}(z_1, z_2, z_3) \sim\ & -2\pi(\varepsilon_1 y_{z_1} + \varepsilon_2 y_{z_2} + \varepsilon_3 y_{z_3}) \\
& +\alpha(y_{z_3})(x_{z_3} - \pi) - \alpha(-y_{z_1})(x_{1-z_1} - \pi) \\
& +\alpha(y_{z_1})(x_{z_1} - \pi) - \alpha(y_{z_1-z_2})(x_{1-z_2+z_1} - \pi) + \alpha(-y_{z_2})(x_{1-z_2} - \pi) \\
& +\alpha(y_{z_2})(x_{z_2} - \pi) - \alpha(y_{z_2-z_3})(x_{1-z_3+z_2} - \pi) + \alpha(-y_{z_3})(x_{1-z_3} - \pi) \\
=\ & \alpha(y_{z_1})(x_{z_1} - \pi - 2\pi\varepsilon_1) + \alpha(-y_{z_1})(x_{z_1} - \pi + 2\pi\varepsilon_1) \\
& +\alpha(y_{z_1-z_2})(x_{z_2-z_1} - \pi) + y_{z_2}(x_{z_2} - \pi - 2\pi\varepsilon_2) \\
& +\alpha(y_{z_2-z_3})(x_{z_3-z_2} - \pi) + y_{z_3}(x_{z_3} - \pi - 2\pi\varepsilon_3)
\end{aligned}
$$

when $|y_{z_1}| + |y_{z_2}| + |y_{z_3}| \to \infty$. If we put $\varepsilon_{23} = \varepsilon_2 + \varepsilon_3$ and $\varepsilon_{123} = \varepsilon_1 + \varepsilon_2 + \varepsilon_3$, the right hand side is further equal to

$$\alpha(y_{z_1})(x_{z_1+z_2+z_3} - 3\pi - 2\pi\varepsilon_{123}) + \alpha(-y_{z_1})(x_{z_1-z_2-z_3} + \pi + 2\pi\varepsilon_{123})$$

along the line $y_{z_2-z_1} = y_{z_3-z_2} = 0$,

$$\alpha(y_{z_2})(x_{z_2+z_3} - 2\pi - 2\pi\varepsilon_{23}) + \alpha(-y_{z_2})(x_{-z_1-z_3} + \pi + 2\pi\varepsilon_{23})$$

along the line $y_{z_1} = y_{z_3-z_2} = 0$, and

$$\alpha(y_{z_3})(x_{z_3} - \pi - 2\pi\varepsilon_3) + \alpha(-y_{z_3})(x_{-z_2} + 2\pi\varepsilon_3)$$

along the line $y_{z_1} = y_{z_2-z_1} = 0$. Therefore, $(\xi_1, \xi_2, \xi_3) \in \Delta_{\varepsilon_1,\varepsilon_2,\varepsilon_3}$ must obey

$$\xi_1 + \xi_2 + \xi_3 - \frac{3}{2} - \varepsilon_{123} < 0, \quad \xi_1 - \xi_2 - \xi_3 + \frac{1}{2} + \varepsilon_{123} < 0,$$

$$\xi_2 + \xi_3 - 1 - \varepsilon_{23} < 0, \quad -\xi_1 - \xi_3 + \frac{1}{2} + \varepsilon_{23} < 0,$$

$$\xi_3 - \frac{1}{2} - \varepsilon_3 < 0, \quad -\xi_2 + \xi_3 < 0,$$

and so we can observe $\Delta_{\varepsilon_1,\varepsilon_2,\varepsilon_3} = \emptyset$ unless $\varepsilon_1 = \varepsilon_2 = \varepsilon_3 = 0$. In fact, we have

$$\varepsilon_3 = 0, \quad \xi_3 < \frac{1}{2}$$

from the fifth and the sixth inequalities first,

$$\varepsilon_{23} = 0, \quad \frac{1}{2} - (\xi_1 + \xi_3) < 0 < 1 - (\xi_2 + \xi_3)$$

from the third and the fourth inequalities second, and

$$\varepsilon_{123} = 0, \quad \frac{1}{2} - (-\xi_1 + \xi_2 + \xi_3) < 0 < \frac{3}{2} - (\xi_1 + \xi_2 + \xi_3)$$

from the first and the second inequalities.

On the other hand, $Im\mathscr{L}(1 - z_{\nu+1} + z_\nu)$ is smooth at $(z_1, z_2, z_3) = (\xi_1, \xi_2, \xi_3)$ if and only if $\xi_\nu < \xi_{\nu+1}$, and so $\Delta_{0,0,0}$ is the set of $(\xi_1, \xi_2, \xi_3) \in [0, 1]^3$ satisfying

$$\frac{1}{2} < -\xi_1 + \xi_2 + \xi_3 < 1 < \xi_1 + \xi_2 + \xi_3 < \frac{3}{2},$$

$$\frac{1}{2} < \xi_1 + \xi_3 < \xi_2 + \xi_3 < 2\xi_3 < 1.$$

In what follows, we use a natural projection $p : \mathbb{C}^n \to \mathbb{R}^n$ defined by

$$p(z_1, \ldots, z_n) = (\operatorname{Re} z_1, \ldots, \operatorname{Re} z_n).$$

Then, for each $(\xi_1, \ldots, \xi_n) \in \Delta_{\varepsilon_1, \ldots, \varepsilon_n}$, the function

$$(f_{\varepsilon_1, \ldots, \varepsilon_n})|_{p^{-1}(\xi_1, \ldots, \xi_n)} : p^{-1}(\xi_1, \ldots, \xi_n) \to \mathbb{R}$$

must have a global minimum by definition. Therefore, there exist a decomposition

$$\Delta_{\varepsilon_1, \ldots, \varepsilon_n} = \Delta^{(1)}_{\varepsilon_1, \ldots, \varepsilon_n} \cup \cdots \cup \Delta^{(m)}_{\varepsilon_1, \ldots, \varepsilon_n}$$

and smooth sections

$$s^{(\mu)}_{\varepsilon_1, \ldots, \varepsilon_n} : \Delta^{(\mu)}_{\varepsilon_1, \ldots, \varepsilon_n} \to p^{-1}(\Delta^{(\mu)}_{\varepsilon_1, \ldots, \varepsilon_n}),$$

where $1 \leq \mu \leq m$, such that, for $(\xi_1, \ldots, \xi_n) \in \Delta^{(\mu)}_{\varepsilon_1, \ldots, \varepsilon_n}$, the function $(f_{\varepsilon_1, \ldots, \varepsilon_n})|_{p^{-1}(\xi_1, \ldots, \xi_n)}$ has a global minimum at $s^{(\mu)}_{\varepsilon_1, \ldots, \varepsilon_n}(\xi_1, \ldots, \xi_n)$. Then, any critical point of

$$f_{\varepsilon_1, \ldots, \varepsilon_n} \circ s^{(\mu)}_{\varepsilon_1, \ldots, \varepsilon_n} : \Delta^{(\mu)}_{\varepsilon_1, \ldots, \varepsilon_n} \to \mathbb{R}$$

is a critical point of $H_{\varepsilon_1, \ldots, \varepsilon_n}$, which must be a local maximum of $f_{\varepsilon_1, \ldots, \varepsilon_n} \circ s^{(\mu)}_{\varepsilon_1, \ldots, \varepsilon_n}$ by the Cauchy-Riemann equation. Without loss of generality, we can suppose

$$f_{\varepsilon_1, \ldots, \varepsilon_n} \circ s^{(1)}_{\varepsilon_1, \ldots, \varepsilon_n}, \ldots, f_{\varepsilon_1, \ldots, \varepsilon_n} \circ s^{(l)}_{\varepsilon_1, \ldots, \varepsilon_n}$$

have local maxima, where $f_{\varepsilon_1,\dots,\varepsilon_n} \circ s^{(1)}_{\varepsilon_1,\dots,\varepsilon_n}$ has the largest one at a critical point $\zeta_{\varepsilon_1,\dots,\varepsilon_n}$ of $H_{\varepsilon_1,\dots,\varepsilon_n}$, and

$$f_{\varepsilon_1,\dots,\varepsilon_n} \circ s^{(l+1)}_{\varepsilon_1,\dots,\varepsilon_n}, \dots, f_{\varepsilon_1,\dots,\varepsilon_n} \circ s^{(m)}_{\varepsilon_1,\dots,\varepsilon_n}$$

have no local maxima. Then, we define

$$\Omega_{\varepsilon_1,\dots,\varepsilon_n} = \begin{cases} \Delta^{(1)}_{\varepsilon_1,\dots,\varepsilon_n} & \text{if } l \geq 1, \\ \emptyset & \text{if } l = 0. \end{cases}$$

If $\Omega_{\varepsilon_1,\dots,\varepsilon_n} \neq \emptyset$, we denote $s^{(1)}_{\varepsilon_1,\dots,\varepsilon_n}$ by $s_{\varepsilon_1,\dots,\varepsilon_n}$.

Example 4.5 Suppose K is the twist knot with 6 crossings. Then, we have

$$\frac{\partial H}{\partial z_1} = 2\pi\sqrt{-1}\left\{-\log\left(1 - \frac{1}{Q_1}\right) + \log\left(1 - \frac{Q_1}{Q_2}\right) - \log(1 - Q_1)\right\},$$

$$\frac{\partial H}{\partial z_2} = 2\pi\sqrt{-1}\left\{-\log\left(1 - \frac{Q_1}{Q_2}\right) + \log\left(1 - \frac{1}{Q_2}\right) - \log(1 - Q_2)\right.$$

$$\left. + \log\left(1 - \frac{Q_2}{Q_3}\right)\right\},$$

$$\frac{\partial H}{\partial z_3} = 2\pi\sqrt{-1}\left\{-\log(1 - Q_3) - \log\left(1 - \frac{Q_2}{Q_3}\right) + \log\left(1 - \frac{1}{Q_3}\right)\right\}.$$

If we fix $(\xi_1, \xi_2, \xi_3) \in \Delta_{0,0,0}$, a critical point (η_1, η_2, η_3) of $(f_{0,0,0})|_{p^{-1}(\xi_1,\xi_2,\xi_3)}$ must satisfy

$$0 = \arg(1 - 1/Q_1) - \arg(1 - Q_1/Q_2) + \arg(1 - Q_1),$$

$$0 = \arg(1 - 1Q_1/Q_2) - \arg(1 - 1/Q_2) + \arg(1 - Q_2) - \arg(1 - Q_2/Q_3),$$

$$0 = \arg(1 - Q_3) + \arg(1 - Q_2/Q_3) - \arg(1 - 1/Q_3),$$

which is equivalent to

$$\frac{(1 - 1/Q_1)(1 - Q_1)}{1 - Q_1/Q_2}, \ \frac{(1 - Q_1/Q_2)(1 - Q_2)}{(1 - 1/Q_2)(1 - Q_2/Q_3)}, \ \frac{(1 - Q_3)(1 - Q_2/Q_3)}{1 - 1/Q_3} > 0,$$

that is,

$$(1 - 1/Q_1)(1 - Q_1)Q_2Q_3, \ Q_3(Q_2 - Q_1), \ Q_2 - Q_3 > 0.$$

Therefore, (η_1, η_2, η_3) is a solution to

$$0 = \text{Im}\,(1 - 1/Q_1)(1 - Q_1)Q_2 Q_3$$
$$= 2e^{-2\pi(\eta_2+\eta_3)} X\{\cos\pi(\xi_1 + \xi_2 + \xi_3) - e^{2\pi\eta_1}\cos\pi(-\xi_1 + \xi_2 + \xi_3)\},$$
$$0 = \text{Im}\,Q_3(Q_2 - Q_1) = e^{-2\pi(\eta_2+\eta_3)}\sin 2\pi(\xi_2 + \xi_3) - e^{-2\pi(\eta_1+\eta_3)}\sin 2\pi(\xi_1 + \xi_3),$$
$$0 = \text{Im}\,(Q_2 - Q_3) = e^{-2\pi\eta_2}\sin 2\pi\xi_2 - e^{-2\pi\eta_3}\sin 2\pi\xi_3,$$

where we put

$$X = -e^{-2\pi\eta_1}\sin\pi(\xi_1 + \xi_2 + \xi_3) + \sin\pi(-\xi_1 + \xi_2 + \xi_3).$$

On the other hand, $\text{Re}\,(1 - 1/Q_1)(1 - Q_1)Q_2 Q_3$ is equal to

$$2e^{-2\pi(\eta_2+\eta_3)}(e^{-2\pi\eta_1}X^2 - e^{-2\pi\eta_1} + 2\cos 2\pi\xi_1 - e^{2\pi\eta_1}),$$

which is negative if $X = 0$. Thus, $(f_{0,0,0})|_{p^{-1}(\xi_1,\xi_2,\xi_3)}$ has the global minimum at

$$\eta_1 = \frac{1}{2\pi}\log\frac{\cos\pi(\xi_1 + \xi_2 + \xi_3)}{\cos\pi(-\xi_1 + \xi_2 + \xi_3)},$$
$$\eta_2 = \frac{1}{2\pi}\log\frac{\sin 2\pi(\xi_2 + \xi_3)\cos\pi(\xi_1 + \xi_2 + \xi_3)}{\sin 2\pi(\xi_1 + \xi_3)\cos\pi(-\xi_1 + \xi_2 + \xi_3)},$$
$$\eta_3 = \frac{1}{2\pi}\log\frac{\sin 2\pi\xi_3\sin 2\pi(\xi_2 + \xi_3)\cos\pi(\xi_1 + \xi_2 + \xi_3)}{\sin 2\pi\xi_2\sin 2\pi(\xi_1 + \xi_3)\cos\pi(-\xi_1 + \xi_2 + \xi_3)}$$

for each $(\xi_1, \xi_2, \xi_3) \in \Delta_{0,0,0}$ and we can observe $\Omega_{0,0,0} = \Delta_{0,0,0}$ in this case.

In what follows, by Λ, we denote the $n \times n$ matrices

$$\left(\frac{\partial^2 H_{\varepsilon_1,\ldots,\varepsilon_n}}{\partial z_i \partial z_j}\right)$$

evaluated at $\zeta_{\varepsilon_1,\ldots,\varepsilon_n}$. The goal of this subsection is the following.

Lemma 4.2 *Suppose $\Omega_{\varepsilon_1,\ldots,\varepsilon_n} \neq \emptyset$. Suppose also that there exists a homotopy*

$$h : \partial\Omega_{\varepsilon_1,\ldots,\varepsilon_n} \times [0, 1] \to E$$

between the identity and $s_{\varepsilon_1,\ldots,\varepsilon_n}|_{\partial\Omega_{\varepsilon_1,\ldots,\varepsilon_n}}$, where

$$E = f_{\varepsilon_1,\ldots,\varepsilon_n}^{-1}((-\infty, f_{\varepsilon_1,\ldots,\varepsilon_n}(\zeta_{\varepsilon_1,\ldots,\varepsilon_n}))) \cap p^{-1}(\Omega_{\varepsilon_1,\ldots,\varepsilon_n}).$$

Then, if Λ is non-singular,

$$\langle K \rangle_N^{\Omega_{\varepsilon_1,\dots,\varepsilon_n}} = N^{\frac{3}{2}} \exp \frac{N H_{\varepsilon_1,\dots,\varepsilon_n}(\zeta_{\varepsilon_1,\dots,\varepsilon_n})}{2\pi\sqrt{-1}} \cdot O(1)$$

when N is large.

Proof Let Φ_t be the flow on $p^{-1}(\Omega_{\varepsilon_1,\dots,\varepsilon_n})$ generated by the gradient vector field of $f_{\varepsilon_1,\dots,\varepsilon_n}$, that is,

$$\frac{d}{dt}\Phi_t(z_1,\dots,z_n) = \left(\frac{\partial f_{\varepsilon_1,\dots,\varepsilon_n}}{\partial \bar{z}_1}, \dots, \frac{\partial f_{\varepsilon_1,\dots,\varepsilon_n}}{\partial \bar{z}_n} \right),$$

and put

$$I_{\varepsilon_1,\dots,\varepsilon_n} = \{(z_1,\dots,z_n) \in p^{-1}(\Omega_{\varepsilon_1,\dots,\varepsilon_n}) : \lim_{t\to\infty}\Phi_t(z_1,\dots,z_n) = \zeta_{\varepsilon_1,\dots,\varepsilon_n}\}.$$

Note that $\operatorname{Re} H_{\varepsilon_1,\dots,\varepsilon_n}$ is constant on $I_{\varepsilon_1,\dots,\varepsilon_n}$ because

$$\frac{\operatorname{Re} H_{\varepsilon_1,\dots,\varepsilon_n}}{dt} = \sum_{\nu=1}^{n} \operatorname{Re}\left\{ \frac{\partial(\operatorname{Re} H_{\varepsilon_1,\dots,\varepsilon_n})}{\partial z_\nu} \cdot \frac{dz_\nu}{dt} \right\}$$

$$= \sum_{\nu=1}^{n} \operatorname{Re}\left\{ \sqrt{-1} \cdot \frac{\partial(\operatorname{Im} H_{\varepsilon_1,\dots,\varepsilon_n})}{\partial z_\nu} \cdot \frac{dz_\nu}{dt} \right\}$$

$$= \sum_{\nu=1}^{n} \operatorname{Re}\left\{ \sqrt{-1} \cdot \frac{\partial f_{\varepsilon_1,\dots,\varepsilon_n}}{\partial z_\nu} \cdot \frac{\partial f_{\varepsilon_1,\dots,\varepsilon_n}}{\partial \bar{z}_\nu} \right\} = 0.$$

Note also that $\zeta_{\varepsilon_1,\dots,\varepsilon_n}$ is a critical point of $f_{\varepsilon_1,\dots,\varepsilon_n}$ whose Morse index is n as Λ is non-singular and that, for small $\epsilon > 0$,

$$E_+ = f_{\varepsilon_1,\dots,\varepsilon_n}^{-1}((-\infty, f_{\varepsilon_1,\dots,\varepsilon_n}(\zeta_{\varepsilon_1,\dots,\varepsilon_n}) + \epsilon]) \cap p^{-1}(\Omega_{\varepsilon_1,\dots,\varepsilon_n})$$

is obtained from

$$E_- = f_{\varepsilon_1,\dots,\varepsilon_n}^{-1}((-\infty, f_{\varepsilon_1,\dots,\varepsilon_n}(\zeta_{\varepsilon_1,\dots,\varepsilon_n}) - \epsilon]) \cap p^{-1}(\Omega_{\varepsilon_1,\dots,\varepsilon_n})$$

by attaching an n-handle W whose core is $\bar{I}_{\varepsilon_1,\dots,\varepsilon_n} \cap W$, where $\bar{I}_{\varepsilon_1,\dots,\varepsilon_n}$ denotes the closure of $I_{\varepsilon_1,\dots,\varepsilon_n}$. As $f|_{\bar{I}_{\varepsilon_1,\dots,\varepsilon_n}}$ is concave while $f|_{p^{-1}(\zeta_{\varepsilon_1,\dots,\varepsilon_n})}$ is convex near $\zeta_{\varepsilon_1,\dots,\varepsilon_n}$, we can suppose that, if $(\xi_1,\dots,\xi_n)$ belongs to

$$U = \{(\xi_1,\dots,\xi_n) \in \Omega_{\varepsilon_1,\dots,\varepsilon_n} : |(\xi_1,\dots,\xi_n) - p(\zeta_{\varepsilon_1,\dots,\varepsilon_n})| < N^{-\gamma}\},$$

where $1/3 < \gamma < 1/2$, $p^{-1}(\xi_1, \ldots, \xi_n)$ intersects $\bar{I}_{\varepsilon_1,\ldots,\varepsilon_n} \cap W$ transversely in one point, say $\sigma_{\varepsilon_1,\ldots,\varepsilon_n}(\xi_1, \ldots, \xi_n)$. Note that, if we put $(\hat{\xi}_1, \ldots, \hat{\xi}_n) = (\xi_1, \ldots, \xi_n) - p(\zeta_{\varepsilon_1,\ldots,\varepsilon_n})$ and

$$
J(\hat{\xi}_1, \ldots, \hat{\xi}_n) = \left(\frac{\sigma_{\varepsilon_1,\ldots,\varepsilon_n}}{\partial \xi_1} \cdots \frac{\sigma_{\varepsilon_1,\ldots,\varepsilon_n}}{\partial \xi_n} \right),
$$

$H_{\varepsilon_1,\ldots,\varepsilon_n}(\sigma_{\varepsilon_1,\ldots,\varepsilon_n}(\xi_1, \ldots, \xi_n))$ is equal to

$$
H_{\varepsilon_1,\ldots,\varepsilon_n}(\zeta_{\varepsilon_1,\ldots,\varepsilon_n}) + (\hat{\xi}_1, \ldots, \hat{\xi}_n) \cdot {}^t J(0) \Lambda J(0) \cdot {}^t(\hat{\xi}_1, \ldots, \hat{\xi}_n) + O(N^{-3\gamma}).
$$

Thus, by putting $(\lambda_{ij}) = {}^t J(0) \Lambda J(0)$ and $(\check{\xi}_1, \ldots, \check{\xi}_n) = N^{\frac{1}{2}}(\hat{\xi}_1, \ldots, \hat{\xi}_n)$, we have

$$
N^{\frac{n+3}{2}} \int_{\sigma_{\varepsilon_1,\ldots,\varepsilon_n}(U)} \Psi_N(z_1, \ldots, z_n) \prod_{\nu=1}^{n} (-Q_\nu^N)^{\varepsilon_\nu} dz_1 \cdots dz_n
$$

$$
= N^{\frac{n+3}{2}} e^{\frac{N}{2\pi\sqrt{-1}} H_{\varepsilon_1,\ldots,\varepsilon_n}(\zeta_{\varepsilon_1,\ldots,\varepsilon_n}) + T(\zeta_{\varepsilon_1,\ldots,\varepsilon_n})} \int_U e^{-\frac{N}{2\pi} \sum_{i,j=1}^{n} \lambda_{ij}\hat{\xi}_i\hat{\xi}_j + O(N^{1-3\gamma})} |J(\hat{\xi}_1, \ldots, \hat{\xi}_n)| d\hat{\xi}_1 \cdots d\hat{\xi}_n
$$

$$
= N^{\frac{3}{2}} e^{\frac{N}{2\pi\sqrt{-1}} H_{\varepsilon_1,\ldots,\varepsilon_n}(\zeta_{\varepsilon_1,\ldots,\varepsilon_n}) + T(\zeta_{\varepsilon_1,\ldots,\varepsilon_n})} |J(0)| \int_{\check{\xi}_1^2 + \cdots + \check{\xi}_n^2 < N^{1-2\gamma}} e^{-\frac{1}{2\pi} \sum_{i,j=1}^{n} \lambda_{ij}\check{\xi}_i\check{\xi}_j + O(N^{1-3\gamma})} d\check{\xi}_1 \cdots d\check{\xi}_n
$$

$$
= N^{\frac{3}{2}} e^{\frac{N}{2\pi\sqrt{-1}} H_{\varepsilon_1,\ldots,\varepsilon_n}(\zeta_{\varepsilon_1,\ldots,\varepsilon_n}) + T(\zeta_{\varepsilon_1,\ldots,\varepsilon_n})} |J(0)| \cdot \sqrt{\frac{(2\pi^2)^n}{|{}^t J(0)\Lambda J(0)|}} \cdot \{1 + O(N^{1-3\gamma})\}.
$$

This completes the proof because $\Omega_{\varepsilon_1,\ldots,\varepsilon_n}$ is homotopic to

$$
h(\partial \Omega_{\varepsilon_1,\ldots,\varepsilon_n} \times [0, 1]) \cup s_{\varepsilon_1,\ldots,\varepsilon_n}(\Omega_{\varepsilon_1,\ldots,\varepsilon_n} \setminus U) \cup g(\partial U \times [0, 1]) \cup \sigma_{\varepsilon_1,\ldots,\varepsilon_n}(U),
$$

where $g : U \times [0, 1] \to p^{-1}(U)$ is a homotopy between $s_{\varepsilon_1,\ldots,\varepsilon_n}$ and $\sigma_{\varepsilon_1,\ldots,\varepsilon_n}$, and

$$
h(\partial \Omega_{\varepsilon_1,\ldots,\varepsilon_n} \times [0, 1]) \cup s'_{\varepsilon_1,\ldots,\varepsilon_n}(\Omega_{\varepsilon_1,\ldots,\varepsilon_n} \setminus U) \cup g(\partial U \times [0, 1])
$$

is contained in E.

Remark 4.1 By using Lemma 4.2, the asymptotic behavior of Kashaev's invariant of the knot 5_2, the twist knot with 5 crossings, is computed in [41]. Lemma 4.2 can be also applied to the invariant of the knot 6_1, the twist knot with 6 crossings, but cannot be applied to the invariants for the other twist knots because the values of the potential functions at $\partial[0, 1]^n$ exceed the critical values. On the other hand, in [71], Ohtsuki computed the asymptotic expansion of Kashaev's invariant of the knot 5_2 more precisely by using the Poisson summation formula, which is generalized to the invariants of the knots with 6 crossings in [72].

4.2.4 Remaining Tasks

From now on, we suppose K is hyperbolic and denote the complement of K by M. First of all, we have to show

Conjecture 4.1 There exists $(\varepsilon_1, \ldots, \varepsilon_n) \in \mathbb{Z}^n$ such that $\Omega_{\varepsilon_1, \ldots, \varepsilon_n} \neq \emptyset$.

In what follows, we assume Conjecture 4.1. Note that the second assumption in Lemma 4.2 can be confirmed if $\partial \Omega_{\varepsilon_1, \ldots, \varepsilon_n} \subset E$ and $f_{\varepsilon_1, \ldots, \varepsilon_n}|_{p^{-1}(\xi_1, \ldots, \xi_n)}$ has a unique critical point on $p^{-1}(\xi_1, \ldots, \xi_n)$ for each $(\xi_1, \ldots, \xi_n) \in \partial \Omega_{\varepsilon_1, \ldots, \varepsilon_n}$ as in Example 4.5. However, this condition is too strong, and we had better to show the following in general.

Conjecture 4.2 There exist an extension $\hat{s}_{\varepsilon_1, \ldots, \varepsilon_n} : [0, 1]^n \to E$ of $s_{\varepsilon_1, \ldots, \varepsilon_n}$ and a homotopy

$$\hat{h} : \partial [0, 1]^n \times [0, 1] \to p^{-1}([0, 1]^n)$$

between the identity and $\hat{s}_{\varepsilon_1, \ldots, \varepsilon_n}|_{\partial [0,1]^n}$ such that

$$\lim_{N \to \infty} \frac{2\pi}{N} \cdot \log \left| \int_{\hat{h}(\partial [0,1]^n \times [0,1])} \Psi_N(z_1, \ldots, z_n) \prod_{\nu=1}^{n} (-Q_\nu^N)^{\varepsilon_\nu} dz_1 \cdots dz_n \right|$$

is less than $f_{\varepsilon_1, \ldots, \varepsilon_n}(\zeta_{\varepsilon_1, \ldots, \varepsilon_n})$.

On the other hand, we are interested in the geometrical meaning of $f_{\varepsilon_1, \ldots, \varepsilon_n}(\zeta_{\varepsilon_1, \ldots, \varepsilon_n})$. Recall that, for a representation

$$\rho : \pi_1(M) \to \mathrm{PSL}(2; \mathbb{C}) \simeq \mathrm{Isom}^+ \mathbb{H}^3,$$

the volume $\mathrm{Vol}(\rho)$ of ρ is defined as the volume of the image of the fundamental domain under the ρ-equivariant map from the universal cover of M to $\mathbb{H}^3$.

Conjecture 4.3 There exists a representation $\rho_{\varepsilon_1, \ldots, \varepsilon_n} : \pi_1(M) \to \mathrm{PSL}(2; \mathbb{C})$ such that

$$\mathrm{Vol}(\rho_{\varepsilon_1, \ldots, \varepsilon_n}) = f_{\varepsilon_1, \ldots, \varepsilon_n}(\zeta_{\varepsilon_1, \ldots, \varepsilon_n}).$$

Conjecture 4.3 is true at least for alternating knots. In the next section, we will explain how to prove it. Note that, if Conjectures 4.3 is true, the non-singularity of $\Lambda_{\varepsilon_1, \ldots, \varepsilon_n}$ follows from the existence of the deformation of $\rho_{\varepsilon_1, \ldots, \varepsilon_n}$ and that, if Conjecture 4.2 is also true, we have

$$\langle K; \varepsilon_1, \ldots, \varepsilon_n \rangle_N \sim N^{\frac{3}{2}} \exp \frac{N H_{\varepsilon_1, \ldots, \varepsilon_n}(\zeta_{\varepsilon_1, \ldots, \varepsilon_n})}{2\pi \sqrt{-1}}$$

by Lemma 4.2. Therefore, to prove the volume conjecture, we further need the following.

Conjecture 4.4 There exists $(\varepsilon_1, \ldots, \varepsilon_n) \in \mathbb{Z}^n$ such that $\rho_{\varepsilon_1, \ldots, \varepsilon_n}$ is discrete and faithful.

4.3   Geometric Part

In this section, following [79, 92], we observe how the complement M of a hyperbolic knot K in S^3 decomposes into ideal tetrahedra which correspond to the q-factorials in Kashaev's invariant of K, and relate the potential function of K to the hyperbolicity equations for the triangulation, the hyperbolic volume of M, and the other geometric invariants. Good references for such ideal triangulations of cusped hyperbolic three-manifolds are [70] and [86].

Recall that the 3-dimensional hyperbolic space $\mathbb{H}^3$ is the upper half space of $\mathbb{R}^3$ endowed with the metric

$$ds^2 = \frac{dx^2 + dy^2 + dz^2}{z^2},$$

where the group of orientation preserving isometries is PSL$(2; \mathbb{C})$. Then, a knot K in S^3 is said to be *hyperbolic* if there is a discrete, torsion-free subgroup Γ of PSL$(2; \mathbb{C})$ such that the complement M of K is homeomorphic to $\mathbb{H}^3/\Gamma$.

4.3.1   Ideal Triangulation

A tetrahedron in $\mathbb{H}^3$ whose four vertices are placed in $\partial \mathbb{H}^3 = \mathbb{C} \cup \{\infty\}$ is called *ideal*. See Fig. 4.4. The shape of such a tetrahedron is determined by a complex number z associated to a pair of opposite edges, called *modulus*, and its volume is given by the Bloch-Wigner function

Fig. 4.4 Moduli associated to the vertical edges

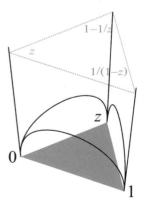

Fig. 4.5 The octahedron
between the overpass and the
underpass

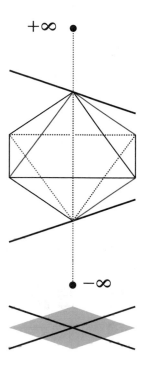

$$\mathscr{D}(z) = \operatorname{Im} \operatorname{Li}_2(z) + \log |z| \arg(1 - z).$$

For simplicity, we suppose K has an alternating diagram D. First, we put an octahedron between the overpass and the underpass of K corresponding to each crossing of D as shown in Fig. 4.5, which is divided into four tetrahedra around the vertical axis referred as *crossing edge*. Next, we glue the red edges at the vertical edge between K and the north pole $+\infty$, and glue the blue edges at the vertical edge between K and the south pole $-\infty$, which makes each tetrahedron as shown in Fig. 4.6. In what follows, the edge of each tetrahedron connecting the two poles, which is opposite to the crossing edge and corresponds to a face of D, is referred as *face edge*. Finally, we can glue the octahedra corresponding to the ends of each edge of D as shown in Fig. 4.7, and we obtain an ideal triangulation $\mathscr{T}$ of $S^3 \backslash (K \cup \{\pm\infty\})$ which is not hyperbolic however.

Let e be an edge of D and Θ_e the intersection of the octahedra corresponding to the ends of e. Then, we obtain an ideal triangulation $\mathscr{S}$ of M from $\mathscr{T}$ by collapsing Θ_e to a point on K, where the tetrahedra of $\mathscr{T}$ touching Θ_e are collapsed. First, the tetrahedra corresponding to the four corners around e, which intersect Θ_e in 2-cells, degenerate to crossing edges in $\mathscr{S}$. Second, the tetrahedra corresponding to the corners in the two faces of D incident to e, which intersect Θ_e in face edges, degenerate to triangles in $\mathscr{S}$. Third, the tetrahedra corresponding to the corners around the edge located at the "entrance" and "exit" of D, which intersect Θ_e in 1-cells other than face edges, degenerate to triangles in $\mathscr{S}$. See Fig. 4.8.

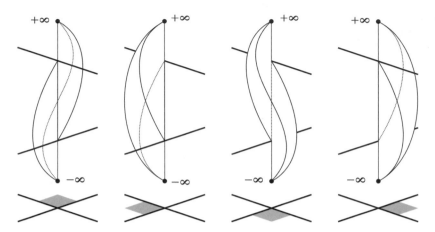

Fig. 4.6 Four tetrahedra after gluing

Fig. 4.7 The edges with the
same color are identified

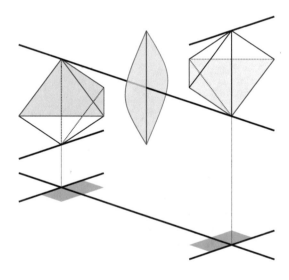

Fig. 4.8 Tetrahedra
corresponding to the corners
with stars are collapsed

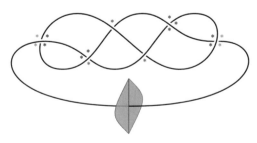

It should be noted that the corners of D corresponding to the tetrahedra collapsed above exactly coincide with the corners of D corresponding to the q-factorials which are eliminated algebraically when we compute Kashaev's invariant from the $(1,1)$-tangle presentation of K obtained from D by cutting e.

4.3.2 Cusp Triangulation

In what follows, we denote the regular neighborhoods of K, $+\infty$, $-\infty$, $\partial \Theta$, $K \cup \Theta$ in S^3 by

$$N(K), \; N(+\infty), \; N(-\infty), \; N(\partial \Theta), \; N(K \cup \Theta)$$

respectively. Let $S_1, \ldots, S_m$ denote the tetrahedra in $\mathscr{S}$ and $w_1, \ldots, w_n$ their moduli associated to crossing edges, and so to face edges, which we assign to the corners of D corresponding to $S_1, \ldots, S_m$. In this subsection, to understand the ideal triangulation $\mathscr{S}$ of M, we are going to write down the triangulation of $\partial N(K \cup \Theta)$ induced by $\mathscr{S}$ explicitly.

Notice that, before the collapsing, each octahedron intersects $\partial N(K)$ in two "rings" and intersects $\partial N(\pm \infty)$ in two "bows" as shown in Fig. 4.9. Therefore, the triangulation $\partial_0 \mathscr{T}$ of $\partial N(K)$ induced by $\mathscr{T}$ can be obtained by gathering these rings. See Fig. 4.10, where we assign w_μ to each cut-end of S_μ near the cut-ends of its crossing edge in the right hand side. Similarly, the triangulation $\partial_+ \mathscr{T}$ of $\partial N(+\infty)$ induced by $\mathscr{T}$ can be obtained by gathering the upper bows, and the triangulation $\partial_- \mathscr{T}$ of $\partial N(-\infty)$ induced by $\mathscr{T}$ can be obtained by gathering the lower bows. See Figs. 4.11 and 4.12, where we assign w_μ to each cut-end of S_μ near the cut-end of its face edge in the right hand side.

Eventually, $\partial_+ \mathscr{T}$ and $\partial_- \mathscr{T}$ are obtained from D and its mirror image by star-subdividing their faces respectively.

Now, we are going to describe the cell decomposition $\partial \mathscr{T}$ of $\partial N(K \cup \Theta)$ induced by $\mathscr{T}$ which will be deformed to the triangulation of $\partial N(\partial \Theta)$ induced by $\mathscr{S}$.

First, the cell decomposition $\partial_\pm \mathscr{T}$ of the torus $\partial N(\partial \Theta)$ induced by $\mathscr{T}$ is easily obtained by tubing $\partial_+ \mathscr{T}$ and $\partial_+ \mathscr{T}$ with the boundary of the tubular neighborhood of $\partial \Theta$ in $S^3 \setminus N(\pm \infty)$ as shown in Fig. 4.13. Second, the cell decomposition $\partial \mathscr{T}$ of the torus $\partial N(K \cup \Theta)$ induced by $\mathscr{T}$ is obtained by gluing $\partial_0 \mathscr{T}$ and $\partial_\pm \mathscr{T}$ along the two ends of the bicollar neighborhood of Θ in $S^3 \setminus \{N(K) \cup N(\partial \Theta)\}$ as shown in Fig. 4.14.

Note that $\partial \mathscr{T}$ consists of a cell decomposition of an annulus corresponding to $\partial N(\partial \Theta)$, the left hand side of Fig. 4.14, and a cell decomposition of an annulus corresponding to $\partial N(K)$, the right hand side of Fig. 4.14.

Finally, the triangulation $\partial \mathscr{S}$ of $\partial N(K \cup \Theta)$ induced by $\mathscr{S}$ can be obtained from $\partial \mathscr{T}$ by collapsing the cut-ends of the degenerate tetrahedra as indicated in Fig. 4.14.

Fig. 4.9 Bows and rings

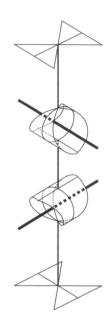

4.3.3   Hyperbolicity Equations

If the hyperbolic structures of $S_1, \ldots, S_m$ determined by $w_1, \ldots, w_m$ fit together, the product of the moduli around each edge of $\mathscr{S}$ should be 1, which is called the *edge relation*. Furthermore, if the structure of M is complete, the product of the moduli along any meridian of K should be 1, which is called the *cusp condition*. Such conditions are called the *hyperbolicity equations* for M. Curious to say, such equations are related to the potential function which appears in the integral expression of Kashaev's invariant.

Theorem 4.1 *The hyperbolicity equations for M can be obtained from the potential function $H(z_1, \ldots, z_n)$, that is,*

$$\exp\left\{\frac{1}{2\pi\sqrt{-1}}\frac{\partial H}{\partial z_1}\right\} = 1, \ \ldots, \ \exp\left\{\frac{1}{2\pi\sqrt{-1}}\frac{\partial H}{\partial z_n}\right\} = 1,$$

where the moduli $w_1, \ldots, w_m$ are given as the ratios of z_ν's.

The following example describes the proof.

Example 4.6 Let K be the knot depicted in Fig. 4.15. Then, as in Example 4.3, the potential function is given by

$$H(z_1, z_2, z_3) = -\mathscr{L}(1 - z_1) + \mathscr{L}(z_1) - \mathscr{L}(z_1 - z_2) + \mathscr{L}(1 - z_2)$$
$$- \mathscr{L}(z_2) + \mathscr{L}(z_2 - z_3) - \mathscr{L}(1 - z_3) + \mathscr{L}(z_3).$$

Fig. 4.10 A development of
$\partial N(K)$, where each
horizontal line represents a
meridian of K and each
vertical line represents a
longitude of K

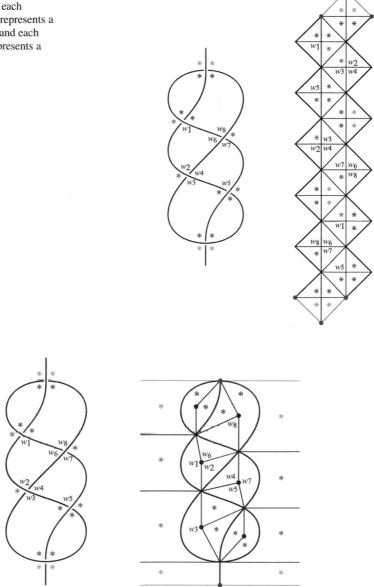

Fig. 4.11 A development of $\partial N(+\infty)$ with two vertices corresponding to the unbounded regions
of D removed, where the edges at the top and the bottom are identified

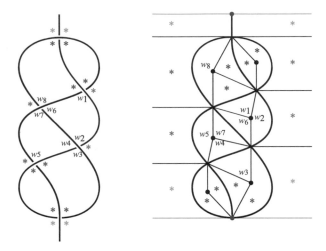

Fig. 4.12 A development of $\partial N(-\infty)$ with two vertices corresponding to the unbounded regions of D removed, where the edges at the top and the bottom are identified

Fig. 4.13 A development of $\partial N(\partial \Theta)$, where the edges at the top and the bottom are identified

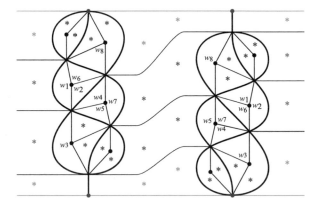

Then, the edge relations around the face edges can be easily read from the left hand side of Fig. 4.16, that is,

$$w_1 w_2 w_6 = 1 = w_4 w_5 w_7.$$

Similarly, the edge relations around the crossing edges can be easily read from the right hand side of Fig. 4.16, that is,

$$w_2 w_3 w_4 = 1 = w_6 w_7 w_8.$$

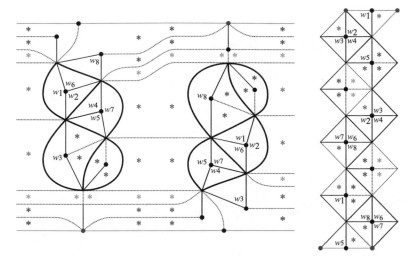

Fig. 4.14 A development of $\partial N(K \cup \Theta)$, where the dotted edges are contracted

Fig. 4.15 The variables of
the potential function are
assigned to the edges of D

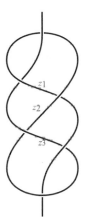

The cusp conditions along two meridians between the both sides of Fig. 4.16 can be
read easily, that is,

$$\frac{w_1}{w_8} = 1 = \frac{w_3}{w_5}.$$

These equations suggest to write the moduli $w_1, \ldots, w_8$ as the ratios of the variables
Q_1, Q_2, Q_3 corresponding to the edges of D assigned z_1, z_2, z_3, that is,

$$w_1 = Q_1, \ w_8 = Q_1, \ w_6 = \frac{Q_2}{Q_1}, \ w_7 = \frac{1}{Q_2}, \ w_3 = Q_3, \ w_4 = \frac{Q_2}{Q_1}, \ w_2 = \frac{1}{Q_2},$$

$$w_5 = Q_3.$$

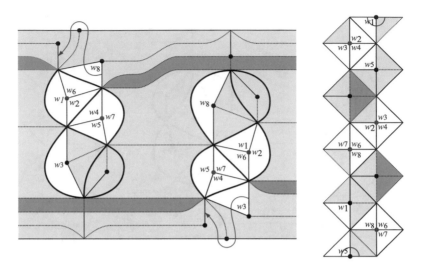

Fig. 4.16 Simple cusp conditions

Fig. 4.17 Essential equations

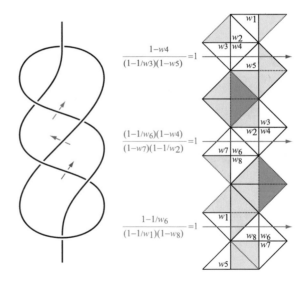

The other equations can be read from the right hand side of Fig. 4.16. In fact, by taking the product of the moduli along the annuli indicated in Fig. 4.17, which corresponds to the edges of D assigned z_1, z_2, z_3, we have

$$1 = \frac{1 - 1/w_6}{(1 - 1/w_1)(1 - w_8)} = \frac{1 - Q_1/Q_2}{(1 - 1/Q_1)(1 - Q_1)},$$

Fig. 4.18 Fundamental
domain of M

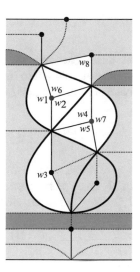

$$1 = \frac{(1 - 1/w_6)(1 - w_4)}{(1 - w_7)(1 - 1/w_2)} = \frac{(1 - Q_1/Q_2)(1 - Q_2/Q_1)}{(1 - 1/Q_2)(1 - Q_2)},$$

$$1 = \frac{1 - w_4}{(1 - 1/w_3)(1 - w_5)} = \frac{1 - Q_2/Q_1}{(1 - 1/Q_1)(1 - Q_3)}.$$

The right hand sides exactly coincide with

$$\exp\left\{\frac{1}{2\pi\sqrt{-1}}\frac{\partial H}{\partial z_1}\right\}, \quad \exp\left\{\frac{1}{2\pi\sqrt{-1}}\frac{\partial H}{\partial z_2}\right\}, \quad \exp\left\{\frac{1}{2\pi\sqrt{-1}}\frac{\partial H}{\partial z_3}\right\}$$

if we put $Q_\nu = e^{2\pi\sqrt{-1}z_\nu}$.

Remark 4.2 Figure 4.18, a part of Fig. 4.16, gives a nice view of the fundamental domain of M in $\mathbb{H}^3$ from ∞. If the edges of $\mathscr{S}$ are homotopically non-trivial, each solution to the equations in Theorem 4.1, such as $\zeta_{\varepsilon_1,\ldots,\varepsilon_n}$ in the previous section, determines the shape of this domain and a holonomy representation of $\pi_1(M)$ into PSL(2; $\mathbb{C}$). In particular, there must exists a solution corresponding to the discrete faithful representation.

If D is alternating, the edges of $\mathscr{S}$ is known to be essential [27, 78], and Conjecture 4.3 is true by Theorem 4.2 below. However, we cannot confirm Conjecture 4.4 because the solution corresponding to the discrete faithful representation may not coincide with $\zeta_{\varepsilon_1,\ldots,\varepsilon_n}$.

4.3.4 Complex Volumes

In this subsection, we study the critical value of the potential function at the geometric solution to the equations in Theorem 4.1. Let $\mathrm{Vol}(M)$ be the *hyperbolic volume* of M and $\mathrm{cs}(M)$ the *Chern–Simons invariant* of M, that is,

$$\mathrm{cs}(M) = \frac{1}{8\pi^2} \int_{s(M)} \mathrm{tr}(A \wedge dA + A \wedge A \wedge A) \in \mathbb{R}/\mathbb{Z}, \tag{4.4}$$

where A and s denote the connection and a section of the orthonormal frame bundle of M. Since M is a cusped hyperbolic three-manifold, $\mathrm{cs}(M)$ is only defined modulo $1/2$. See [58] for detail. Then, we define the *complex volume* of M by

$$\mathrm{cv}(M) = -2\pi^2 \,\mathrm{cs}(M) + \sqrt{-1}\,\mathrm{Vol}(M) \mod \pi^2. \tag{4.5}$$

Theorem 4.2 *Suppose that $(z_1, \ldots, z_n) = (\zeta_1, \ldots, \zeta_n)$ is the geometric solution to the hyperbolicity equations and that*

$$\frac{1}{2\pi\sqrt{-1}} \frac{\partial H}{\partial z_v} = 2\pi\sqrt{-1} \cdot \varepsilon_v,$$

where $\varepsilon_v \in \mathbb{Z}$. Then, we have

$$\mathrm{cv}(M) \equiv H(\zeta_1, \ldots, \zeta_n) - 2\pi\sqrt{-1}\sum_{v=1}^{n} \varepsilon_v \log \zeta_v \mod \pi^2.$$

Proof We prove the imaginary part of the equality only. See [94] for the proof of the real part. From the definition of Bloch-Wigner function, we can observe that

$$\mathrm{Im}\,\mathrm{Li}_2(z/w) = \mathscr{D}(z/w) + \log|z| \cdot \mathrm{Im}\left\{ z \frac{\partial\,\mathrm{Li}_2(z/w)}{\partial z} \right\}$$

$$+ \log|w| \cdot \mathrm{Im}\left\{ w \frac{\partial\,\mathrm{Li}_2(z/w)}{\partial w} \right\}.$$

and that

$$\mathrm{Im}\,H(z_1, \ldots, z_n) = \sum_{\mu=1}^{m} \mathscr{D}(w_\mu) + \sum_{v=1}^{n} \log|z_v| \cdot \mathrm{Im}\left\{ \frac{1}{2\pi\sqrt{-1}} \frac{\partial H}{\partial z_v} \right\}.$$

Since $\mathscr{D}(w_\mu)$ gives the volume of S_μ at $(z_1, \ldots, z_n) = (\zeta_1, \ldots, \zeta_n)$, we have

$$\operatorname{Im} H(\zeta_1, \ldots, \zeta_n) = \operatorname{Vol}(M) + \sum_{\nu=1}^{n} 2\pi \varepsilon_\nu \log |\zeta_\nu|.$$

This proves the imaginary part of the equality.

Remark 4.3 Theorem 4.2 can be generalized to a formula for the complex volumes of the representations corresponding to the other solutions. See [97] for the Chern–Simons invariant of representations.

Chapter 5
Representations of a Knot Group, Their Chern–Simons Invariants, and Their Reidemeister Torsions

Abstract In this chapter, we describe representations of the fundamental group of a knot complement to SL$(2; \mathbb{C})$ by giving examples. We also give the definitions of the Chern–Simons invariant and the Reidemeister torsion associated with such a representation. We also give examples of calculation. We will explain relations of these invariants to the asymptotic behavior of the colored Jones polynomial in the next chapter.

5.1 Representations of a Knot Group

5.1.1 Presentation

We consider the complement of a knot $S^3 \setminus K$ in the three-sphere S^3. Let us denote by $\pi_1(K)$ the *fundamental group* $\pi_1(S^3 \setminus K; x_0)$ with appropriate basepoint x_0.[1] We sometimes call $\pi_1(K)$ the *knot group*.

Given a knot diagram we can present the knot group in the following way. As in Fig. 5.1, we assign a loop with endpoints at ∞ to each arc of the diagram. Here the diagram is cut by crossings into several segments, each of them is called an arc. Then around a positive crossing we have the relation

$$yxz^{-1}x^{-1} = 1$$

as we can read off from Fig. 5.2. Similarly around a negative crossing we have the relation

$$xyx^{-1}z^{-1} = 1$$

---

[1] Usually we regard S^3 as the one-point compactification of $\mathbb{R}^3$, that is $S^3 = \mathbb{R}^3 \cup \{\infty\}$, and put $x_0 := \infty$.

© The Author(s), under exclusive licence to Springer Nature Singapore Pte Ltd. 2018
H. Murakami, Y. Yokota, *Volume Conjecture for Knots*, SpringerBriefs in Mathematical Physics 30, https://doi.org/10.1007/978-981-13-1150-5_5

Fig. 5.1 Generator

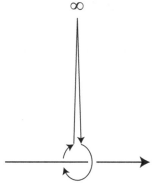

Fig. 5.2 Relation at a
positive crossing

Fig. 5.3 Relation at a
negative crossing

as we can see from Fig. 5.3. So if there are n crossings in the diagram we have the
following presentation for $\pi_1(K)$:

$$\langle x_1, x_2, \ldots, x_n \mid r_1, r_2, \ldots, r_n \rangle,$$

where the r_i are relations as above. Note that one relation, say r_n, is unnecessary
because a small loop around a crossing can be shrunk through the point at infinity
in S^2. Therefore we have the following presentation for $\pi_1(K)$.

Definition 5.1 (Wirtinger presentation) If a knot K has a diagram with n crossings, we have the following presentation for $\pi_1(K)$.

$$\pi_1(K) = \langle x_1, x_2, \ldots, x_n \mid r_1, r_2, \ldots, r_{n-1} \rangle.$$

We call it the Wirtinger presentation.

Let $N(K) \subset S^3$ be the tubular neighborhood of K in S^3 and $\partial N(K)$ be its
boundary. Note that $N(K)$ is homeomorphic to a solid torus $D^2 \times S^1$ and that $\partial N(K)$
is homeomorphic to a torus $S^1 \times S^1$.

Definition 5.2 (median) A simple closed curve μ on $\partial N(K)$ that is nullhomotopic in $N(K)$ and oriented so that $\mathrm{lk}(\mu, K) = 1$ is called the meridian, where lk denotes the linking number.

Choose a path γ connecting ∞ and a point in μ. We also call the homotopy class $[\gamma \mu \gamma^{-1}] \in \pi_1(K)$ the meridian of K. Note that the meridian is defined up to conjugation.

Definition 5.3 (longitude) A simple closed curve λ on $\partial N(K)$ that is parallel to K in $N(K)$ and nullhomologous in $S^3 \setminus \mathrm{Int}\, N(K)$ is called the longitude.[2]

For a path γ connecting ∞ and a point in λ, the homotopy class $[\gamma \lambda \gamma^{-1}] \in \pi_1(K)$ is also called the longitude of K. It is defined up to conjugation.

Example 5.1 (Figure-eight knot) Let x, y, z, and w be the generators of $\pi_1(\mathscr{E})$ indicated in Fig. 5.4. Then we have

$$\pi_1(\mathscr{E}) = \langle x, y, z, w \mid xyx^{-1}z^{-1}, zwz^{-1}x^{-1}, zyw^{-1}y^{-1} \rangle.$$

Note that we do not write the relation around the bottom-right crossing.

From the first relation, we have $z = xyx^{-1}$, and from the second relation we have $w = z^{-1}xz = xy^{-1}xyx^{-1}$. So we have the following presentation with two generators and one relation.

$$\pi_1(\mathscr{E}) = \langle x, y \mid \omega x = y\omega \rangle, \tag{5.1}$$

where we put $\omega := xy^{-1}x^{-1}y$.

We can choose x as the meridian.

Fig. 5.4 Generators of $\pi_1(\mathscr{E})$

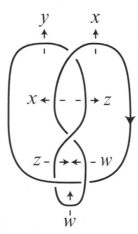

---

[2]It is often called the preferred longitude. Sometimes any simple closed curve on $\partial N(K)$ that is parallel to K in $N(K)$ is also called the longitude. Our (preferred) longitude λ satisfies $\mathrm{lk}(\lambda, K) = 0$.

Fig. 5.5 Generators of
$\pi_1(T(2, 2a + 1))$

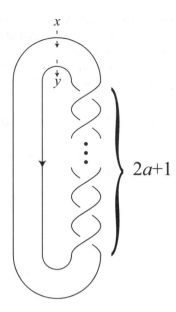

Choosing a simple closed curve from the top-right along the knot, the longitude is

$$wx^{-1}yz^{-1} = xy^{-1}xyx^{-2}yxy^{-1}x^{-1}. \qquad (5.2)$$

Example 5.2 (Torus knot of type $(2, 2a + 1)$) It would be a good exercise to check that $\pi_1(T(2, 2a + 1))$ has the following presentation:

$$\pi_1(T(2, 2a + 1)) = \langle x, y \mid (xy)^a x = y(xy)^a \rangle,$$

where x and y are generators indicated in Fig. 5.5. We can choose x as the meridian and the longitude is

$$(xy)^a xy^{-a}(xy)^a x^{-3a-1} = y(xy)^{2a} x^{-4a-1} \qquad (5.3)$$

if we choose a simple closed curve from the top-right. Note that we need to multiply by x^{-3a-1} to make the linking number 0.

5.1.2   Representation

A homomorphism $\rho: \pi_1(K) \to \mathrm{SL}(2; \mathbb{C})$ is called a *representation* to $\mathrm{SL}(2; \mathbb{C})$. In this book a representation to $\mathrm{SL}(2; \mathbb{C})$ is simply called a representation.

Definition 5.4 (Abelian representation) A representation ρ is called Abelian if $\mathrm{Im}(\rho)$ forms an Abelian subgroup of $SL(2; \mathbb{C})$. A representation that is not Abelian is called non-Abelian.

Since $H_1(S^3 \setminus K) \cong \mathbb{Z}$ and it is generated by the meridian, the Abelianization of $\pi_1(K)$ is $\mathbb{Z}$ generated by the median $\alpha([\mu])$, where $\alpha \colon \pi_1(K) \to H_1(S^3 \setminus K)$ is the Abelianization map. Therefore the image of an Abelian representation is generated by the image of the meridian.

So an Abelian representation is either ρ_z^A sending the meridian to $\begin{pmatrix} z & 0 \\ 0 & z^{-1} \end{pmatrix}$ ($z \neq$ 0) or $\rho_{0,\pm}^A$ sending the meridian to $\pm \begin{pmatrix} 1 & 1 \\ 0 & 1 \end{pmatrix}$, up to conjugation. Note that ρ_z^A is conjugate to $\rho_{z^{-1}}^A$.

Definition 5.5 (Reducible representation) A representation ρ is called reducible if there exists a one-dimensional subspace $V \subset \mathbb{C}^2$ such that $\rho(g)(V) \subset V$ for any $g \in \pi_1(K)$. Note that this is equivalent to say that there exists a non-singular 2×2 matrix S such that $S^{-1}\rho(g)S$ is upper-triangular for any $g \in \pi_1(K)$. A representation that is not reducible is called *irreducible*.

Definition 5.6 (Affine representation) A reducible, non-Abelian representation is sometimes called an affine representation since it corresponds to the affine transformation $z \mapsto az + b$.

Suppose ρ is such a representation. Up to conjugation we can write the image of x_i as $\begin{pmatrix} p_i & q_i \\ 0 & p_i^{-1} \end{pmatrix}$, where the x_i are generators given in Definition 5.1. Then from the relation r_j, we have $p_i = p_1$ for any i, and

$$(p^2 - 1)q_j - p^2 q_k + q_i = 0 \quad \text{when the crossing is positive,}$$

$$(p^2 - 1)q_j - p^2 q_i + q_k = 0 \quad \text{when the crossing is negative,}$$

where we put $p := p_1$, $x_j := x$, $x_i := y$, and $x_k := z$. This is a system of $(n-1)$ linear equations with n indeterminates. Comparing this with the Alexander matrix $\left(\alpha \left(\frac{\partial r_j}{\partial x_i} \right) \right)_{i=1,2,\dots,n}^{j=1,2,\dots,n-1}$, one can conclude that if p^2 is a zero of the *Alexander polynomial*[3] there is an affine representation. Here $\frac{\partial}{\partial x_i}$ is the Fox derivative[4] and $\alpha \colon \mathbb{Z}[\pi_1(K)] \to \mathbb{Z}[t, t^{-1}]$ is the Abelianization map sending x_i to t. See [13, 18]. See also [51, 2.4.3. Corollary] and [43, Exercise 11.2].

In the following we exhibit irreducible representations for the figure-eight knot and the torus knot of type $(2, 2a + 1)$.

---

[3]The Alexander polynomial is $\det \left(\alpha \left(\frac{\partial r_j}{\partial x_i} \right) \right)_{i=1,2,\dots,n-1}^{j=1,2,\dots,n-1}$.

[4]See Sect. 5.3.1.

Example 5.3 (Figure-eight knot) Since $\pi_1(\mathcal{E})$ has the presentation (5.1), it is enough to give $\rho(x)$ and $\rho(y)$ such that $\rho(\omega)\rho(x) = \rho(y)\rho(\omega)$.

For a complex number m ($m \neq 0$) let $\rho_{m,\pm}$ be the non-Abelian representation defined as follows:

$$\rho_{m,\pm}(x) := \begin{pmatrix} m & 1 \\ 0 & m^{-1} \end{pmatrix},$$

$$\rho_{m,\pm}(y) := \begin{pmatrix} m & 0 \\ d_\pm & m^{-1} \end{pmatrix},$$

(5.4)

where

$$d_\pm = \frac{1}{2}\left(-m^2 - m^{-2} + 3 \pm \sqrt{(m^2 + m^{-2} + 1)(m^2 + m^{-2} - 3)}\right). \qquad (5.5)$$

It can be proved that any non-Abelian representation is conjugate to $\rho_{m,\pm}$ for some m [75]. Note that $\rho_{m^{-1},\pm}$ is conjugate to $\rho_{m,\pm}$. Note also that $\rho_{m,+}$ and $\rho_{m,-}$ are not conjugate if $d_\pm \neq 0$, because

$$\rho_{m,\pm}(xy) = \begin{pmatrix} m^2 + d_\pm & m^{-1} \\ d_\pm m^{-1} & m^{-2} \end{pmatrix}$$

and they have different traces.

The longitude (5.2) is sent to

$$\begin{pmatrix} \ell(m)^{\pm 1} & \mp(m + m^{-1})\sqrt{(m^2 + m^{-2} + 1)(m^2 + m^{-2} - 3)} \\ 0 & \ell(m)^{\mp 1} \end{pmatrix},$$

where

$$\ell(m) := \frac{\left(m^4 - m^2 - 2 - m^{-2} + m^{-4}\right)}{2} - \frac{(m^2 - m^{-2})}{2}$$
$$\sqrt{(m^2 + m^{-2} + 1)(m^2 + m^{-2} - 3)}. \qquad (5.6)$$

Note that $\ell(m)$ is a solution to the following equation.

$$\ell - (m^4 - m^2 - 2 - m^{-2} + m^{-4}) + \ell^{-1} = 0, \qquad (5.7)$$

which coincides with the A-polynomial of the figure-eight knot [17].

If $d_\pm = 0$, then $\rho_{m,\pm}$ becomes reducible, giving affine representations. They are given by

$$\rho_\pm^{\text{affine}}(x) := \begin{pmatrix} (\sqrt{5}+1)/2 & 1 \\ 0 & (\sqrt{5}-1)/2 \end{pmatrix},$$

$$\rho_\pm^{\text{affine}}(y) := \begin{pmatrix} (\sqrt{5}+1)/2 & 0 \\ 0 & (\sqrt{5}-1)/2 \end{pmatrix}$$

when $m = (\sqrt{5}+1)/2$ and

$$\rho_\pm^{\text{affine}}(x) := \begin{pmatrix} -(\sqrt{5}+1)/2 & 1 \\ 0 & -(\sqrt{5}-1)/2 \end{pmatrix},$$

$$\rho_\pm^{\text{affine}}(y) := \begin{pmatrix} -(\sqrt{5}+1)/2 & 0 \\ 0 & -(\sqrt{5}-1)/2 \end{pmatrix}$$

when $m = -(\sqrt{5}+1)/2$.

When $m = 1$, $\rho_{1,\pm}$ gives the holonomy representations. The images of the meridian and the longitude are $\begin{pmatrix} 1 & 1 \\ 0 & 1 \end{pmatrix}$ and $\begin{pmatrix} -1 & \mp 2\sqrt{3}\sqrt{-1} \\ 0 & -1 \end{pmatrix}$, respectively. If we regard $\mathrm{SL}(2;\mathbb{C})$ (precisely speaking $\mathrm{PSL}(2;\mathbb{C})$) as the group of orientation preserving isometries of the upper half hyperbolic space $\mathbb{H}^3$, it gives translations $z \mapsto z+1$ and $z \mapsto z \pm 2\sqrt{3}\sqrt{-1}$ on $\partial \mathbb{H}^3 = \mathbb{C}$. Since the pair $(1, 2\sqrt{3}\sqrt{-1})$ gives the positive orientation in $\mathbb{H}^3$, the representation $\rho_{1,+}$ gives the complete hyperbolic structure with positive volume and $\rho_{1,-}$ gives the one with negative volume.

Example 5.4 (Torus knot of type $(2, 2a+1)$) For a complex number m ($m \neq 0$) and an integer j ($0 \leq j \leq a-1$), put

$$\rho_{m,j}(x) := \begin{pmatrix} m & 1 \\ 0 & m^{-1} \end{pmatrix},$$

$$\rho_{m,j}(y) := \begin{pmatrix} m & 0 \\ 2\cos\left(\frac{(2j+1)\pi}{2a+1}\right) - m^2 - m^{-2} & m^{-1} \end{pmatrix}.$$

Then $\rho_{m,j}$ is a non-Abelian representation. It is known that any non-Abelian representation is conjugate to $\rho_{m,j}$ for some m and j [75]. Note that $\rho_{m^{-1},j}$ is conjugate to $\rho_{m,j}$. Since $\operatorname{Tr}\rho_{m,j}(xy) = 2\cos\left(\frac{(2j+1)\pi}{2a+1}\right)$, $\rho_{m,j}$ is not conjugate to $\rho_{m,j'}$ if $j \neq j'$.

The longitude (5.3) is sent to

$$\begin{pmatrix} -m^{-2(2a+1)} & \frac{m^{2(2a+1)}-m^{-2(2a+1)}}{m-m^{-1}} \\ 0 & -m^{2(2a+1)} \end{pmatrix}.$$ (5.8)

Note that $l = -m^{-2(2a+1)}$, that is, $lm^{2(2a+1)} + 1 = 0$ is the A-polynomial of $T(2, 2a + 1)$.

If $m = \pm \exp\left(\frac{(2j+1)\pi\sqrt{-1}}{2(2a+1)}\right)$, then the representation becomes reducible and defines an affine representation.

5.2   The Chern–Simons Invariant

5.2.1   Definition

Let W be a oriented closed three-manifold and $\rho\colon \pi_1(W) \to SL(2; \mathbb{C})$ be a representation.

Then there is a flat connection A on $W \times SL(2; \mathbb{C})$ that induces ρ as the holonomy representation. Here a flat connection is an $sl_2(\mathbb{C})$-valued 1-form on W satisfying $dA + A \wedge A = 0$. The $SL(2; \mathbb{C})$ *Chern–Simons function* cs_W is defined by

$$cs_W([\rho]) := \frac{1}{8\pi^2} \int_W \mathrm{tr}\left(A \wedge dA + \frac{2}{3} A \wedge A \wedge A\right) \in \mathbb{C} \pmod{\mathbb{Z}},$$ (5.9)

where $[\rho]$ is the conjugacy class. It is known that the representations modulo conjugation is in one-one correspondence to the flat connections modulo gauge equivalence.

If W is hyperbolic, ρ_0 is the holonomy representation associated with the complete hyperbolic metric, and $cs(W)$ is the Chern–Simons invariant defined by using the Levi–Civita connection (see (4.4)), then T. Yoshida [95, Lemma 3.1] proved

$$cs_W([\rho_0]) = cs(W) - \frac{\sqrt{-1}}{\pi^2} \mathrm{Vol}(W).$$

See also [47, P. 554]. Therefore the complex volume $cv(W)$ defined in (4.5) coincides with $-\pi^2 cs_W([\rho_0])$ in this case.

In [47], P. Kirk and E. Klassen introduced the Chern–Simons invariant for a three-manifold whose boundary consists of tori. They also gave a way to calculate the Chern–Simons invariant of a three-manifold obtained by pasting two such manifolds.

Let M be an oriented three-manifold with boundary ∂M. For simplicity we assume that ∂M is a torus. The SL(2; $\mathbb{C}$) character variety $X(L)$ of a manifold L is the set of the traces of representations from $\pi_1(L)$ to SL(2; $\mathbb{C}$).

First, we study the character variety $X(\partial M)$. Consider a map $p\colon \mathrm{Hom}(\pi_1(\partial M), \mathbb{C}) \to X(\partial M)$ defined by

$$p(\kappa) := \left[\gamma \mapsto \begin{pmatrix} e^{2\pi \sqrt{-1}\kappa(\gamma)} & 0 \\ 0 & e^{-2\pi\sqrt{-1}\kappa(\gamma)} \end{pmatrix} \right]$$

for $\gamma \in \pi_1(\partial M)$, where the square brackets mean the equivalence class in $X(\partial M)$.

Now fix a generator system (μ, λ) of $\pi_1(\partial M) \cong \mathbb{Z} \oplus \mathbb{Z}$, and take its dual basis (μ^*, λ^*) of $\mathrm{Hom}(\pi_1(\partial M), \mathbb{C})$. Then we see that p is invariant under the following actions on $\mathrm{Hom}(\pi_1(\partial M); \mathbb{C})$:

$$x \cdot (\alpha, \beta) := (\alpha + 1, \beta),$$

$$y \cdot (\alpha, \beta) := (\alpha, \beta + 1),$$

$$b \cdot (\alpha, \beta) := (-\alpha, -\beta),$$

where (α, β) is the element $\alpha\mu^* + \beta\lambda^* \in \mathrm{Hom}(\partial M, \mathbb{C})$. These actions form the group

$$G := \langle x, y, b \mid xyx^{-1}y^{-1} = bxbx = byby = b^2 = 1 \rangle.$$

Note that the quotient space $\mathrm{Hom}(\pi_1(\partial M), \mathbb{C})/G$ is identified with the SL(2; $\mathbb{C}$) character variety $X(\partial M)$.

Let G act on $\mathrm{Hom}(\pi_1(\partial M), \mathbb{C}) \times \mathbb{C}^*$ in the following way.

$$x \cdot (\alpha, \beta; z) := (\alpha + 1, \beta; z\exp(-8\pi\sqrt{-1}\beta)),$$

$$y \cdot (\alpha, \beta; z) := (\alpha, \beta + 1; z\exp(8\pi\sqrt{-1}\alpha)),$$

$$b \cdot (\alpha, \beta; z) := (-\alpha, -\beta; z).$$

We denote the quotient space $(\mathrm{Hom}(\pi_1(\partial M), \mathbb{C}) \times \mathbb{C}^*)/G$ by $E(\partial M)$. So if we denote by $[\alpha, \beta; z]$ the representative, then we have

$$\begin{aligned} [\alpha, \beta; z] &= [\alpha + 1, \beta; z\exp(-8\pi\sqrt{-1}\beta)] \\ &= [\alpha, \beta + 1; z\exp(8\pi\sqrt{-1}\alpha)] \qquad (5.10) \\ &= [-\alpha, -\beta; z]. \end{aligned}$$

Then $q\colon E(\partial M) \to X(\partial M)$ ($q\colon [\alpha, \beta; z] \mapsto [\alpha, \beta]$) becomes a $\mathbb{C}^*$-bundle, where $[\alpha, \beta; z]$ ($[\alpha, \beta]$, respectively) is the equivalence class of $(\alpha, \beta; z)$ $((\alpha, \beta)$, respectively) in $E(\partial M)$ ($X(\partial M)$, respectively).

The Chern–Simons function c_M is a map from $X(M)$ to $E(\partial M)$ such that the following diagram commutes, where i^* is the restriction map.

$$
\begin{array}{ccc}
 & & E(\partial M) \\
 & {\scriptstyle c_M}\nearrow & \downarrow {\scriptstyle q} \\
X(M) & \xrightarrow[\;i^*\;]{} & X(\partial M)
\end{array}
$$

Suppose that an oriented closed three-manifold W is given as $M_1 \cup M_2$, where M_1 and M_2 are manifolds with torus boundaries. We give the same basis for $\pi_1(\partial M_1)$ and $\pi_1(-\partial M_2)$. For a representation ρ of $\pi_1(W)$, let ρ_i be its restriction to $\pi_1(M_i)$ ($i = 1, 2$). Then we have

$$
e^{2\pi\sqrt{-1}\,\mathrm{cs}_W\big([\rho]\big)} = z_1 z_2,
$$

where $c_{M_i}([\rho_i]) = [\alpha, \beta, z_i]$ with respect to the common basis [47, Theorem 2.2].

If $M = S^3 \setminus \mathrm{Int}\, N(K)$ for a knot K, then we can define the $SL(2; \mathbb{C})$ Chern–Simons invariant as follows.

Let μ and λ be the meridian and the longitude, respectively. Up to a conjugation we may assume that a given representation ρ satisfies the following:

$$
\rho(\mu) = \begin{pmatrix} e^{u/2} & * \\ 0 & e^{-u/2} \end{pmatrix} \quad \text{and} \quad \rho(\lambda) = \begin{pmatrix} -e^{v/2} & * \\ 0 & -e^{-v/2} \end{pmatrix}
$$

because μ and λ commute. Note the minus signs in $\rho(\lambda)$. We put these signs because it is known that for a hyperbolic knot the trace of the image of the longitude by the holonomy representation is -2 (see [15]).

Then the $SL(2; \mathbb{C})$ Chern–Simons function c_M has the following form.

$$
c_M\big([\rho]\big) = \left[\frac{u}{4\pi\sqrt{-1}}, \frac{v}{4\pi\sqrt{-1}}; \exp\left(\frac{2}{\sqrt{-1}\pi}\,\mathrm{CS}_{u,v}([\rho])\right)\right]. \tag{5.11}
$$

We call $\mathrm{CS}_{u,v}([\rho])$ the $SL(2; \mathbb{C})$ *Chern–Simons invariant* of $[\rho]$ associated with (u, v). Note that $\mathrm{CS}_{u,v}([\rho])$ is defined modulo $\pi^2\mathbb{Z}$. Note also that $\mathrm{CS}_{0,0}([\rho_0]) = \mathrm{cv}(M)$ for a hyperbolic three-manifold M with ρ_0 the holonomy representation.

5.2.2 How to Calculate

The following theorem is useful to calculate the Chern–Simons invariant.

Theorem 5.1 (Kirk–Klassen's theorem [47]) Let $\rho_t \colon \pi_1(M) \to \mathrm{SL}(2; \mathbb{C})$ be a path of representations $(0 \le t \le 1)$. *We assume that the images of μ and λ have the following form.*

$$\rho_t(\mu) = \begin{pmatrix} e^{u_t/2} & * \\ 0 & e^{-u_t/2} \end{pmatrix},$$

$$\rho_t(\lambda) = \begin{pmatrix} -e^{v_t/2} & * \\ 0 & -e^{-v_t/2} \end{pmatrix}$$

up to conjugation.
  Suppose that c_M is given as

$$c_M([\rho_t]) = \left[\frac{u_t}{4\pi\sqrt{-1}}, \frac{v_t}{4\pi\sqrt{-1}}; z_t \right].$$

Then we have

$$\frac{z_1}{z_0} = \exp\left[\frac{\sqrt{-1}}{2\pi} \int_0^1 \left(u_t \frac{d\,v_t}{d\,t} - v_t \frac{d\,u_t}{d\,t} \right) dt \right].$$

Example 5.5 (Hyperbolic knot) Let $\mathcal{H}$ be a hyperbolic knot and put $M_{\mathcal{H}} := S^3 \setminus \mathrm{Int}\, N(\mathcal{H})$, where $N(\mathcal{H})$ is the regular neighborhood of $\mathcal{H}$. Let $\rho_0 \colon \pi_1(M_{\mathcal{H}}) = \pi_1(\mathcal{H}) \to \mathrm{SL}(2; \mathbb{C})$ be the representation associated with the complete hyperbolic structure. We can deform the complete structure of $S^3 \setminus \mathcal{H}$ by a small complex parameter u. Let ρ_u be the representation associated with u. So ρ_u determines an incomplete hyperbolic structure if $u \ne 0$. By conjugation we assume

$$\rho(\mu) = \begin{pmatrix} e^{u/2} & * \\ 0 & e^{-u/2} \end{pmatrix},$$

$$\rho(\lambda) = \begin{pmatrix} -e^{v(u)/2} & * \\ 0 & -e^{-v(u)/2} \end{pmatrix},$$

where μ and λ are the meridian and the longitude of $\pi_1(\mathcal{H})$, respectively. We choose $v(u)$ so that $v(0) = 0$. See for example [70].
  We assume that ρ_{ut} $(0 < t \le 1)$ defines an incomplete hyperbolic structure. We can write

$$c_{M_{\mathcal{H}}}([\rho_{ut}]) = \left[\frac{ut}{4\pi\sqrt{-1}}, \frac{v(ut)}{4\pi\sqrt{-1}}; z_t \right]$$

for some $z_t \ne 0$. From Theorem 5.1 we have

$$\frac{z_1}{z_0} = \exp\left[\frac{\sqrt{-1}}{2\pi} \int_0^1 \left(ut\frac{d\,v(ut)}{dt} - v(ut)\frac{d\,(ut)}{dt}\right)dt\right]$$

$$= \exp\left[\frac{\sqrt{-1}}{2\pi}\left(\Big[ut\,v(ut)\Big]_0^1 - 2u\int_0^1 v(ut)\,dt\right)\right]$$

$$= \exp\left[\frac{\sqrt{-1}}{2\pi}\left(uv(u) - 2\int_0^u v(s)\,ds\right)\right].$$

Since $z_0 = \exp\left[\frac{2}{\pi\sqrt{-1}}\,\mathrm{cv}(M_{\mathcal{H}})\right]$, where $\mathrm{cv}(M_{\mathcal{H}})$ is the hyperbolic volume defined in (4.5), we have

$$\mathrm{CS}_{u,v(u)}([\rho_u]) = \mathrm{cv}(M_{\mathcal{H}}) + \frac{1}{2}\int_0^u v(s)\,ds - \frac{1}{4}uv(u). \tag{5.12}$$

Example 5.6 (Figure-eight knot) We study the figure-eight knot $\mathcal{E}$ more precisely. Since $\ell(1) = -1$, we need to define $v(u)$ as

$$v(u) = 2\log\ell(e^{u/2}) - 2\pi\sqrt{-1} \tag{5.13}$$

so that $v(0) = 0$ (see Example 5.3). We know that $\mathrm{cv}(M_{\mathcal{E}}) = \sqrt{-1}\,\mathrm{Vol}(S^3 \setminus \mathcal{E})$ and so we have

$$\mathrm{CS}_{u,v(u)}([\rho_u]) = \sqrt{-1}\,\mathrm{Vol}\left(S^3 \setminus \mathcal{E}\right) + \int_0^u \log\ell(e^{s/2})\,ds - \frac{1}{2}u\log\ell(e^{u/2})$$

$$- \frac{1}{2}u\pi\sqrt{-1}$$

$$\tag{5.14}$$

from (5.12).

Next we show another way to calculate the Chern–Simons invariant.

Put $\kappa := \log((3 + \sqrt{5})/2)$ and assume that u is a real number with $0 \le u \le \kappa$. Let α_t $(0 \le t \le 1)$ be a path of Abelian representations defined by

$$\alpha_t(x) = \alpha_t(y) := \begin{pmatrix} e^{\kappa t/2} & 0 \\ 0 & e^{-\kappa t/2} \end{pmatrix}$$

and β_t $(0 \le t \le 1)$ be a path of non-Abelian representations defined by

$$\begin{cases} \beta_t(x) := \begin{pmatrix} e^{u_t/2} & 1 \\ 0 & e^{-u_t/2} \end{pmatrix}, \\[2mm] \beta_t(y) := \begin{pmatrix} e^{u_t/2} & 0 \\ d_+\big|_{m:=e^{u_t/2}} & e^{-u_t/2} \end{pmatrix}, \end{cases}$$

where we put $u_t := (1 - t)\kappa + tu$. Then the paths of representations α_t and β_t have the following properties:

(i) α_0 is trivial,
(ii) α_1 and β_0 share the same trace because β_0 is upper-triangular, and
(iii) $\beta_1 = \rho_{e^{u/2},+}$.

From (i) we know that $c_M([\alpha_0])$ is trivial and from (ii) we have $c_M([\alpha_1]) = c_M([\beta_0])$.

We regard x as the meridian μ and the longitude λ is given in (5.3). From Theorem 5.1 we can write

$$\begin{cases} c_M([\alpha_t]) := \left[\dfrac{\kappa t}{4\pi\sqrt{-1}}, 0; w_t\right], \\[4mm] c_M([\beta_t]) := \left[\dfrac{u_t}{4\pi\sqrt{-1}}, \dfrac{2\log \ell(e^{u_t/2}) - 2\pi\sqrt{-1}}{4\pi\sqrt{-1}}; z_t\right], \end{cases}$$

since

$$\begin{cases} \alpha_t(\lambda) = \begin{pmatrix} 1 & 0 \\ 0 & 1 \end{pmatrix}, \\[4mm] \beta_t(\lambda) = \begin{pmatrix} \ell(e^{u_t/2}) & * \\ 0 & \ell(e^{u_t/2})^{-1} \end{pmatrix} \end{cases}$$

from Example 5.3 (see also (5.13)). Then Kirk–Klassen's theorem (Theorem 5.1) shows that $\dfrac{w_1}{w_0} = 1$ and

$$\frac{z_1}{z_0} = \exp\left(\frac{\sqrt{-1}}{2\pi}\int_0^1 \left(u_t \times \frac{d\left(2\log \ell(e^{u_t/2}) - 2\pi\sqrt{-1}\right)}{dt} - (u-\kappa)(2\log \ell(e^{u_t/2}) - 2\pi\sqrt{-1})\right) dt\right).$$

Since $c_M([\alpha_1]) = c_M([\beta_0])$ from Property (ii) and $w_1 = w_0 = 1$ from Property (i), we have

$$\left[\frac{\kappa}{4\pi\sqrt{-1}}, 0; 1\right] = \left[\frac{\kappa}{4\pi\sqrt{-1}}, \frac{2\log \ell(e^{\kappa/2}) - 2\pi\sqrt{-1}}{4\pi\sqrt{-1}}; z_0\right]$$

$$= \left[\frac{\kappa}{4\pi\sqrt{-1}}, -\frac{1}{2}; z_0\right]$$

since $\ell(e^{\kappa/2}) = 1$. However, from the equivalence relation (5.10),[5] we have

$$\left[\frac{\kappa}{4\pi\sqrt{-1}}, 0; 1\right] \approx \left[\frac{\kappa}{4\pi\sqrt{-1}}, -\frac{1}{2}; e^{-\kappa}\right].$$

So we have $z_0 = e^{-\kappa}$ and

$$z_1 = \exp\left(-u + \frac{\sqrt{-1}}{\pi}\int_0^1 \left(u_t \times \frac{d\,\log \ell(e^{u_t/2})}{d\,t} - (u - \kappa)\log \ell(e^{u_t/2})\right) dt\right).$$

Therefore we finally have

$$CS_{u,v(u)}([\rho_{e^{u/2},+}])$$

$$= -\frac{1}{2}u\pi\sqrt{-1} - \frac{1}{2}\int_0^1 \left(u_t \times \frac{d\,\log \ell(e^{u_t/2})}{d\,t} - (u - \kappa)\log \ell(e^{u_t/2})\right) dt$$

modulo $\pi^2\mathbb{Z}$. Since we calculate

$$\int_0^1 \left(u_t \times \frac{d\,\log \ell(e^{u_t/2})}{d\,t} - (u - \kappa)\log \ell(e^{u_t/2})\right) dt$$

$$= \left[u_t\,\log \ell(e^{u_t/2})\right]_0^1 - 2(u - \kappa)\int_0^1 \log \ell(e^{u_t/2})\,dt$$

$$= u\,\log \ell(e^{u/2}) - 2\int_\kappa^u \log \ell(e^{s/2})\,ds,$$

where we put $s := u_t = (1 - t)\kappa + tu$ in the integral, we have

$$CS_{u,v(u)}([\rho_{e^{u/2},+}]) = \int_\kappa^u \log \ell(e^{s/2})\,ds - \frac{1}{2}u\pi\sqrt{-1} - \frac{1}{2}u\,\log \ell(e^{u/2}) \qquad (5.15)$$

modulo $\pi^2\mathbb{Z}$.

Using the dilogarithm function $\mathrm{Li}_2(z) := -\int_0^z \frac{\log(1 - x)}{x}\,dx$, we put

$$S(u) := \mathrm{Li}_2(e^{u-\varphi(u)}) - \mathrm{Li}_2(e^{u+\varphi(u)}) - u\varphi(u) \qquad (5.16)$$

---

[5]In (5.10), β can be shifted only by integers but here we use $-1/2$. We would need to work in $\mathrm{PSL}(2;\mathbb{C})$ rather than $\mathrm{SL}(2;\mathbb{C})$.

with

$$\varphi(u) := \log\left(\frac{1}{2}(2\cosh u - 1 - \sqrt{(2\cosh u + 1)(2\cosh u - 3)})\right). \qquad (5.17)$$

Here the square root is taken so that its imaginary part is positive (note that $(2\cosh(u)-1)(2\cosh(u)-3) < 0$ since $0 < u < \log((3+\sqrt{5})/2) = \mathrm{arccosh}(3/2))$ and the branch of log is the usual one, that is, its imaginary part is between $-\pi$ and π. Note that $\varphi(u)$ satisfies

$$2\cosh u = 2\cosh(\varphi(u)) + 1. \qquad (5.18)$$

Note also that $\varphi(u)$ is purely imaginary since $\left|\frac{1}{2}(2\cosh u - 1 - \sqrt{(2\cosh u + 1)(2\cosh u - 3)})\right| = 1$ and that $-\frac{\pi}{3} < \mathrm{Im}\,\varphi(u) < 0$.
  Now we have

$$\frac{d\,S(u)}{du} = \log(1 - e^{u+\varphi(u)}) - \log(1 - e^{u-\varphi(u)}) - \varphi(u),$$

and so we have

$$\exp\left(\frac{d\,S(u)}{du}\right) = \cosh(2u) - \cosh u - 1 - \sinh u\sqrt{(2\cosh u + 1)(2\cosh u - 3)},$$

which coincides with $\ell(e^{u/2})$ from (5.6). Since $\dfrac{d\,S}{du}(0) = \pi\sqrt{-1} = \log\ell(e^{u/2})$, we see that

$$\log\ell(e^{u/2}) = \frac{d\,S(u)}{du}$$

and so we have

$$v(u) = 2\frac{d\,S(u)}{du} - 2\pi\sqrt{-1} \qquad (5.19)$$

from (5.13). We also have from (5.15)

$$CS_{u,v(u)}([\rho_{e^{u/2},+}]) = S(u) - \frac{1}{2}u\pi\sqrt{-1} - \frac{1}{4}u\left(v(u) + 2\pi\sqrt{-1}\right)$$
$$= S(u) - u\pi\sqrt{-1} - \frac{1}{4}uv(u) \qquad (5.20)$$

since $\varphi(\kappa) = 0$.

Example 5.7 (Torus knot of type $(2, 2a + 1)$) Put $M := S^3 \setminus \mathrm{Int}\,N(T(2, 2a + 1))$.
Recall the definitions in Examples 5.2 and 5.4.

Let α_t $(0 \leq t \leq 1)$ be a path of Abelian representations defined by

$$\alpha_t(x) = \alpha_t(y) := \begin{pmatrix} \exp\left(\frac{(2j+1)\pi\sqrt{-1}}{2(2a+1)}t\right) & 0 \\ 0 & \exp\left(-\frac{(2j+1)\pi\sqrt{-1}}{2(2a+1)}t\right) \end{pmatrix}$$

and β_t $(0 \leq t \leq 1)$ be a path of non-Abelian representations defined by

$$\begin{cases} \beta_t(x) & := \begin{pmatrix} e^{u_t/2} & 1 \\ 0 & e^{-u_t/2} \end{pmatrix}, \\ \beta_t(y) & := \begin{pmatrix} e^{u_t/2} & 0 \\ 2\cos\left(\frac{(2j+1)\pi}{2a+1}\right) - 2\cosh u_t & e^{-u_t/2} \end{pmatrix}, \end{cases}$$

where we put $u_t := \dfrac{(1-t)(2j+1)\pi\sqrt{-1}}{2a+1} + tu$ for a complex number u. Then the paths of representations α_t and β_t have the following properties:

(i) α_0 is trivial,
(ii) α_1 and β_0 share the same trace because β_0 is upper-triangular, and
(iii) $\beta_1 = \rho_{e^{u/2},j}$.

From (i) we know that $c_M([\alpha_0])$ is trivial and from (ii) we have $c_M([\alpha_1]) = c_M([\beta_0])$.

We regard x as the meridian μ and the longitude λ is given in (5.3). From Theorem 5.1 we can write

$$\begin{cases} c_M([\alpha_t]) & := \left[\frac{(2j+1)t}{4(2a+1)}, 0; w_t\right], \\ c_M([\beta_t]) & := \left[\frac{u_t}{4\pi\sqrt{-1}}, \frac{-2(2a+1)u_t+4l\pi\sqrt{-1}}{4\pi\sqrt{-1}}; z_t\right] \end{cases}$$

for an integer l, since

$$\begin{cases} \alpha_t(\lambda) & = \begin{pmatrix} 1 & 0 \\ 0 & 1 \end{pmatrix}, \\ \beta_t(\lambda) & = \begin{pmatrix} -e^{-(2a+1)u_t} & \frac{\sinh\left((2a+1)u_t\right)}{\sinh(u_t/2)} \\ 0 & -e^{(2a+1)u_t} \end{pmatrix} \end{cases}$$

from Example 5.4. Then Kirk–Klassen's theorem (Theorem 5.1) shows that $\dfrac{w_1}{w_0} = 1$ and

$$\frac{z_1}{z_0}$$

$$= \exp\left(\frac{\sqrt{-1}}{2\pi}\int_0^1 \left(u_t \times \left(-2(2a+1)\frac{d\,u_t}{d\,t}\right) - (-2(2a+1)u_t + 4l\pi\sqrt{-1}) \times \frac{d\,u_t}{d\,t}\right) dt\right)$$

$$= \exp\left(2l\left(u - \frac{(2j+1)\pi\sqrt{-1}}{2a+1}\right)\right).$$

Since $c_M([\alpha_1]) = c_M([\beta_0])$ from Property (ii) and $w_1 = w_0 = 1$ from Property (i), we have

$$\left[\frac{2j+1}{4(2a+1)}, 0; 1\right] = \left[\frac{u_0}{4\pi\sqrt{-1}}, \frac{-2(2a+1)u_0 + 4l\pi\sqrt{-1}}{4\pi\sqrt{-1}}; z_0\right]$$

$$= \left[\frac{2j+1}{4(2a+1)}, l - \frac{2j+1}{2}; z_0\right].$$

However, from the equivalence relation (5.10), we have

$$\left[\frac{2j+1}{4(2a+1)}, 0; 1\right] \approx \left[\frac{2j+1}{4(2a+1)}, l - \frac{2j+1}{2}; \exp\left(\frac{2(2j+1)(l-j)\pi\sqrt{-1}}{2a+1}\right)\right].$$

So we have

$$z_0 = \exp\left(\frac{2(2j+1)(l-j-1/2)\pi\sqrt{-1}}{2a+1}\right)$$

and

$$z_1 = \exp\left(2lu - \frac{(2j+1)^2\pi\sqrt{-1}}{2a+1}\right).$$

Therefore we finally have

$$\mathrm{CS}_{u,v(u)}([\rho_{e^{u/2},j}]) = lu\pi\sqrt{-1} + \frac{(2j+1)^2\pi^2}{2(2a+1)} \pmod{\pi^2\mathbb{Z}} \tag{5.21}$$

with $v(u) := -2(2a+1)u + 4l\pi\sqrt{-1}$. Note that this depends on the choice of an integer l.

The Chern–Simons invariant of a general torus knot is given in [22, Proposition 4].

5.3 Twisted SL(2; ℂ) Reidemeister Torsion

In this section, we study the Reidemeister torsion associated with a representation. It is defined as the torsion of a certain chain complex twisted by the representation.

5.3.1 Definition

To define the Reidemeister torsion we consider a compact three-manifold $M_K :=$ $S^3 \setminus \text{Int } N(K)$ instead of an open three-manifold $S^3 \setminus K$. Note that $\pi_1(K) = \pi_1(M_K)$ and we have a presentation of it as described in Sect. 5.1.1. Write Π for $\pi_1(K)$ and let $\rho : \Pi \to \text{SL}(2; \mathbb{C})$ be a representation. Let $\langle x_1, x_2, \ldots, x_n \mid r_1, r_2, \ldots, r_{n-1} \rangle$ be a Wirtinger presentation of Π (Definition 5.1).

Let us consider the universal cover $\tilde{M}_K$ of M_K. The chain complex $C_*(\tilde{M}_K)$ with integer coefficients can be regarded as a $\mathbb{Z}[\Pi]$-module by the deck transformation. The Lie algebra $\text{sl}_2(\mathbb{C})$ can also be regarded as a $\mathbb{Z}[\Pi]$-module by using the adjoint action of $\rho(x)$ for $x \in \Pi$. Here the adjoint action $\text{Ad } \rho(x)$ on $\text{sl}_2(\mathbb{C})$ is defined as $\text{Ad } \rho(x)(g) := \rho(x)^{-1} g \rho(x)$ for $g \in \text{sl}_2(\mathbb{C})$. Then we have the following chain complex:

$$C_2(\tilde{M}_K) \otimes_{\mathbb{Z}[\Pi]} \text{sl}_2(\mathbb{C}) \xrightarrow{\partial_2} C_1(\tilde{M}_K) \otimes_{\mathbb{Z}[\Pi]} \text{sl}_2(\mathbb{C}) \xrightarrow{\partial_1} C_0(\tilde{M}_K) \otimes_{\mathbb{Z}[\Pi]} \text{sl}_2(\mathbb{C}).$$

Note that here we regard M_K as a 2-dimensional CW-complex with 2-cells $\{r_1, r_2, \ldots, r_{n-1}\}$, 1-cells $\{x_1, x_2, \ldots, x_n\}$ and a 0-cell p following the Wirtinger presentation above.[6] The associated homology group is denoted by $H_*(M_K; \rho)$.

Let $\mathbf{c}_i := \{c_{i,1}, c_{i,2}, \ldots, c_{i,l_i}\}$ be a basis of $C_i := C_i(\tilde{M}_K) \otimes_{\mathbb{Z}[\Pi]} \text{sl}_2(\mathbb{C})$, $\mathbf{b}_i := \{b_{i,1}, b_{i,2}, \ldots, b_{i,m_i}\}$ be a set of vectors such that $\{\partial_i(b_{i,1}), \partial_i(b_{i,2}), \ldots, \partial_i(b_{i,m_i})\}$ forms a basis of $B_{i-1} := \text{Im } \partial_i$, $\mathbf{h}_i := \{h_{i,1}, h_{i,2}, \ldots, h_{i,n_i}\}$ be a basis of $H_i := H_i(M_K; \rho)$, $\tilde{h}_{i,k}$ be a lift of $h_{i,k}$ in $Z_i := \text{Ker } \partial_i$, and $\tilde{\mathbf{h}}_i := \{\tilde{h}_{i,1}, \tilde{h}_{i,2}, \ldots, \tilde{h}_{i,n_i}\}$. Then $\partial_{i+1}(\mathbf{b}_{i+1}) \cup \tilde{\mathbf{h}}_i \cup \mathbf{b}_i$ forms a basis of C_i (Fig. 5.6). For two bases $\mathbf{u}$ and $\mathbf{v}$ of a vector space W, let $[\mathbf{u} \mid \mathbf{v}]$ be the determinant of the basis change matrix from $\mathbf{u}$ to $\mathbf{v}$. We define the *Reidemeister torsion* as

---

[6]We start with a bouquet with n loops $x_1, x_2, \ldots, x_n$ sharing the point p. Then we attach $n-1$ disks $r_1, r_2, \ldots, r_{n-1}$ according to the words r_j. It is known that this CW-complex is simple homotopy equivalent to M_K.

Fig. 5.6 Chain complex and
its basis

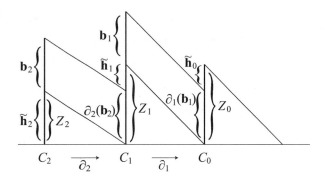

$$\mathrm{Tor}(C_*, \mathbf{c}_*, \mathbf{h}_*) := \prod_{i=0}^{2} \left[\partial_{i+1}(\mathbf{b}_{i+1}) \cup \tilde{\mathbf{h}}_i \cup \mathbf{b}_i \,\middle|\, \mathbf{c}_i \right]^{(-1)^{i+1}}$$

$$= \frac{\left[\partial_2(\mathbf{b}_2) \cup \tilde{\mathbf{h}}_1 \cup \mathbf{b}_1 \,\middle|\, \mathbf{c}_1 \right]}{\left[\partial_1(\mathbf{b}_1) \cup \tilde{\mathbf{h}}_0 \,\middle|\, \mathbf{c}_0 \right] \left[\tilde{\mathbf{h}}_2 \cup \mathbf{b}_2 \,\middle|\, \mathbf{c}_2 \right]} \in \mathbb{C}^*. \tag{5.22}$$

Note that this does not depend on the choices of $\mathbf{b}_i$ nor the choices of lifts of $\mathbf{h}_i$ (see for example [82]). Since the Reidemeister torsion depends only on the simple homotopy type (see [73, p. 10, Remarque (b)], [59]) and the Whitehead group of a knot exterior is trivial [85], we can calculate the torsion regarding M_K as a CW-complex as above.

Following [73, Définition 0.4], we choose geometric bases for C_* as follows. Let $\tilde{p}$ be a lift of p, $\tilde{x}_i$ a lift of x_i ($i = 1, 2, \ldots, n$), and $\tilde{r}_j$ a lift of r_j ($j = 1, 2, \ldots, n-1$). Put $E := \begin{pmatrix} 0 & 1 \\ 0 & 0 \end{pmatrix}$, $H := \begin{pmatrix} 1 & 0 \\ 0 & -1 \end{pmatrix}$ and $F := \begin{pmatrix} 0 & 0 \\ 1 & 0 \end{pmatrix}$. Then $\{E, H, F\}$ is a basis of $\mathrm{sl}_2(\mathbb{C})$. We choose $\{\tilde{p} \otimes E, \tilde{p} \otimes H, \tilde{p} \otimes F\}$ as a geometric basis for C_0, $\{\tilde{x}_1 \otimes E, \tilde{x}_1 \otimes H, \tilde{x}_1 \otimes F, \tilde{x}_2 \otimes E, \tilde{x}_2 \otimes H, \tilde{x}_2 \otimes F, \ldots, \tilde{x}_n \otimes E, \tilde{x}_n \otimes H, \tilde{x}_n \otimes F\}$ as a geometric basis for C_1, and $\{\tilde{r}_1 \otimes E, \tilde{r}_1 \otimes H, \tilde{r}_1 \otimes F, \tilde{r}_2 \otimes E, \tilde{r}_2 \otimes H, \tilde{r}_2 \otimes F, \ldots, \tilde{r}_{n-1} \otimes E, \tilde{r}_{n-1} \otimes H, \tilde{r}_{n-1} \otimes F\}$ as a geometric basis for C_2.

Since the Euler characteristic of M_K is zero, (5.22) does not depend on the choice of geometric bases [73, Définition 0.5].

Definition 5.7 Let ρ be a representation of $\pi_1(K)$. Given a basis $\mathbf{h}_*$ of $H_i(M_K; \rho)$, the torsion $\mathrm{Tor}(C_*, \mathbf{c}_*, \mathbf{h}_*)$ is denoted by $\mathbb{T}^K_{\mathbf{h}_*}(\rho)$. We call it the *twisted Reidemeister torsion* of ρ associated with $\mathbf{h}_*$.

With respect to the geometric bases, the differentials $\partial_2 \colon C_2 \to C_1$ and $\partial_1 \colon C_1 \to C_0$ are given by the Fox free differential calculus (see [53, Chapter 11] for example). Let $\frac{\partial}{\partial x_j}$ be the Fox derivative [24], which is defined by the following rules:

- for words u and v in the x_j, $\frac{\partial (uv)}{\partial x_j} = \frac{\partial u}{\partial x_j} + u \frac{\partial v}{\partial x_j}$,
- for the empty word 1, $\frac{\partial 1}{\partial x_j} = 0$,
- $\frac{\partial x_i}{\partial x_j} = \delta^i_j$, where δ^i_j is Kronecker's delta.

Note that since $0 = \frac{\partial (x \cdot x^{-1})}{\partial x} = 1 + x \cdot \frac{\partial x^{-1}}{\partial x}$, we have $\frac{\partial x^{-1}}{\partial x} = -x^{-1}$.

The differential ∂_2 is given by the following $3(n-1) \times 3n$ matrix:

$$
\partial_2 = \begin{pmatrix} \operatorname{Ad} \rho \left(\frac{\partial r_1}{\partial x_1} \right) & \cdots & \operatorname{Ad} \rho \left(\frac{\partial r_{n-1}}{\partial x_1} \right) \\ \vdots & \ddots & \vdots \\ \operatorname{Ad} \rho \left(\frac{\partial r_1}{\partial x_n} \right) & \cdots & \operatorname{Ad} \rho \left(\frac{\partial r_{n-1}}{\partial x_n} \right) \end{pmatrix},
$$

noting that each $\operatorname{Ad} \rho \left(\frac{\partial r_j}{\partial x_i} \right)$ is a 3×3 matrix. The differential ∂_1 is given by the following $3 \times 3n$ matrix:

$$
\begin{pmatrix} \operatorname{Ad} \rho (x_1 - 1) & \cdots & \operatorname{Ad} \rho (x_n - 1) \end{pmatrix}.
$$

Note that $\partial_1 \circ \partial_2$ is a $3 \times 3(n-1)$ zero matrix, since its $(1, k)$-entry as a 3×3 block matrix is

$$
\operatorname{Ad} \rho (x_1 - 1) \operatorname{Ad} \rho \left(\frac{\partial r_k}{\partial x_1} \right) + \cdots + \operatorname{Ad} \rho (x_n - 1) \operatorname{Ad} \rho \left(\frac{\partial r_k}{\partial x_n} \right),
$$

which vanishes from the fundamental formula of the free differential calculus [24, (2.3)].

Example 5.8 (Abelian representation) We calculate the Reidemeister torsion of the Abelian representation ρ^A_z defined in Definition 5.4.

We denote by $C_2 \xrightarrow{\partial_2} C_1 \xrightarrow{\partial_1} C_0$ the associated chain complex, and by H_i its homology.

Put $X := \operatorname{Ad} \rho^A_z(x_i)$. Then since we have

$$
\rho^A_z(x_i)^{-1} E \rho^A_z(x_i) = \begin{pmatrix} 0 & z^{-2} \\ 0 & 0 \end{pmatrix},
$$

$$
\rho^A_z(x_i)^{-1} H \rho^A_z(x_i) = \begin{pmatrix} 1 & 0 \\ 0 & -1 \end{pmatrix},
$$

$$
\rho^A_z(x_i)^{-1} F \rho^A_z(x_i) = \begin{pmatrix} 0 & 0 \\ z^2 & 0 \end{pmatrix},
$$

X is given by the three by three matrix $\begin{pmatrix} z^{-2} & 0 & 0 \\ 0 & 1 & 0 \\ 0 & 0 & z^2 \end{pmatrix}$ with respect to the basis given before.

Therefore ∂_1 is of the form

$$\partial_1 = \left(\begin{array}{ccc|ccc|ccc} z^{-2} - 1 & 0 & 0 & z^{-2} - 1 & 0 & 0 & & z^{-2} - 1 & 0 & 0 \\ 0 & 0 & 0 & 0 & 0 & 0 & \cdots & 0 & 0 & 0 \\ 0 & 0 & z^2 - 1 & 0 & 0 & z^2 - 1 & & 0 & 0 & z^2 - 1 \end{array} \right).$$

Let $A(t)$ be the $n \times (n-1)$ matrix with (i, j)-entry $\alpha\left(\dfrac{\partial r_j}{\partial x_i} \right)$, where $\alpha \colon \mathbb{Z}[\pi_1(K)] \to \mathbb{Z}[t, t^{-1}]$ is induced by the Abelianization $\pi_1(K) \to \mathbb{Z}$. Then ∂_2 is given by the $3n \times 3(n-1)$ matrix $A(X)$ obtained from $A(t)$ by replacing t with X. It is easy to prove

$$H_2 = \{0\},$$
$$H_1 = \mathbb{C} \quad \text{(generated by } [\tilde{x}_1 \otimes H]),$$
$$H_0 = \mathbb{C} \quad \text{(generated by } [\tilde{p} \otimes H]).$$

Moreover if we put

$$\mathbf{b}_2 := \langle \tilde{r}_1 \otimes E, \tilde{r}_1 \otimes H, \tilde{r}_1 \otimes F, \ldots, \tilde{r}_{n-1} \otimes E, \tilde{r}_{n-1} \otimes H, \tilde{r}_{n-1} \otimes F \rangle,$$
$$\tilde{\mathbf{h}}_1 := \langle \tilde{x}_1 \otimes H \rangle,$$
$$\mathbf{b}_1 := \langle \tilde{x}_1 \otimes E, \tilde{x}_1 \otimes F \rangle,$$
$$\tilde{\mathbf{h}}_0 := \langle \tilde{p} \otimes H \rangle,$$

we see that $\mathbf{b}_2$ forms a basis of C_2, that $\mathbf{b}_1 \cup \tilde{\mathbf{h}}_1 \cup \partial_2(\mathbf{b}_2)$ forms a basis of C_1, and that $\tilde{\mathbf{h}}_0 \cup \partial_1(\mathbf{b}_1)$ forms a basis of C_0.

Therefore the Reidemeister torsion of ρ_z^A associated with these bases is

$$\mathbb{T}_{\mu}^{K}(\rho_{z}^{A}) = \pm \frac{\begin{vmatrix} 1\,0\,0 \\ 0\,0\,1 \\ 0\,1\,0 \\ 0\,0\,0 \\ 0\,0\,0 \\ 0\,0\,0 \quad A(X) \\ \vdots \\ 0\,0\,0 \\ 0\,0\,0 \\ 0\,0\,0 \end{vmatrix}}{\det(I_{3n-3}) \begin{vmatrix} 0 & z^{-2}-1 & 0 \\ 1 & 0 & 0 \\ 0 & 0 & z^{2}-1 \end{vmatrix}}$$

$$= \pm \frac{\det \check{A}(X)}{(z-z^{-1})^{2}},$$

where $\check{A}(X)$ is the $3(n-1) \times 3(n-1)$ matrix obtained from $A(X)$ by removing the first three rows.

Denote by $\check{A}(t)$ the $(n-1) \times (n-1)$ matrix obtained from $A(t)$ by removing the first row. Then the (unnormalized) Alexander polynomial $\tilde{\Delta}(K;t)$ of K is $\det \check{A}(t)$ (up to a multiplication of $\pm t^{k}$). So we have

$$\det \check{A}(X) = \det \check{A}(z^{-2}) \det \check{A}(1) \det \check{A}(z^{2})$$

$$= \tilde{\Delta}(K; z^{-2}) \tilde{\Delta}(K; 1) \tilde{\Delta}(K; z^{2})$$

$$= \Delta(K; z^{2})^{2},$$

where $\Delta(K;t)$ is the normalized Alexander polynomial.[7]

Note that we use $[\tilde{\mu} \otimes H]$ as the generator of H_1 and $[\tilde{p} \otimes H]$ as the generator of H_0, where μ is the meridian of K. Note also that the element $H \in \mathrm{sl}_2(\mathbb{C})$ can be characterized as the vector that is invariant under the adjoint action of $\rho_z^{A}(\mu)$. Since the choices of p and H are natural we use $\mathbb{T}_{\mu}^{K}(\rho_z^{A})$ instead of $T_{\{[\tilde{\mu}\otimes H],[\tilde{p}\otimes H]\}}^{K}(\rho_z^{A})$.

So we prove the following proposition that should be well known to experts.

Proposition 5.1 (Folklore) *The twisted Reidemeister torsion of ρ_z^{A} associated with the meridian is given by*

---

[7]It is normalized so that $\Delta(K; 1) = 1$ and that $\Delta(K; t^{-1}) = \Delta(K; t)$.

$$\mathbb{T}_\mu^K(\rho_z^A) = \pm \left(\frac{\Delta(K; z^2)}{z - z^{-1}} \right)^2.$$

As in the Abelian case above, we need to specify bases $\mathbf{h}_i$ for $i = 0, 1, 2$. We only consider the following case.

Definition 5.8 ([31]) An SL(2; ℂ) representation ρ is called *regular* if it is irreducible and $\dim H_1(M_K; \rho) = 1$.

If a representation ρ is irreducible, then it is easy to prove that $H_0(M_K; \rho) = 0$. So we have $\dim H_2 = \dim H_1 = 1$ for a regular representation since the Euler characteristic of M_K is 0.

So for a regular representation we need to fix bases for H_1 and H_2. To do that we need another regularity.

Definition 5.9 ([73, Définition 3.21] (see also [21])) Let γ be a simple closed curve on ∂M_K. An irreducible representation is called γ-regular if and only if

- The inclusion $\gamma \hookrightarrow M_K$ induces a surjective map $H_1(\gamma; \rho) \twoheadrightarrow H_1(M_K; \rho)$, and
- if $\mathrm{tr}(\rho(\pi_1(\partial M_K))) \subset \{2, -2\}$, then $\rho(\gamma) \neq \pm \begin{pmatrix} 1 & 0 \\ 0 & 1 \end{pmatrix}$.

It can be proved that any γ-regular representation is regular.

We fix an element $P \in \mathrm{sl}_2(\mathbb{C})$ that is invariant under the adjoint action of $\rho(x)$ for any $x \in \pi_1(\partial M_K)$. Then we choose $i_*([\tilde{\gamma}] \otimes P)$ and $i_*\left([\widetilde{\partial M_K}] \otimes P\right)$, called reference generators, as bases of H_1 and H_2 respectively, where $[\gamma] \in H_1(\partial M_K)$ is the homology class of the curve γ, $[\partial M_K] \in H_2(\partial M_K)$ is the fundamental class, and $i \colon \partial M_K \to M_K$ is the inclusion map.

The twisted Reidemeister torsion $\mathbb{T}_\gamma^K(\rho)$ of ρ associated with γ is defined as $\mathrm{Tor}(C_*, \mathbf{c}_*, \mathbf{h}_*)$ defined in (5.22) for these bases [73].

5.3.2 How to Calculate

If one wants to calculate the twisted Reidemeister torsion from the definition, we need to calculate the determinants of huge matrices. In this subsection we will describe an easier way following [21]. See [73], [34, § 4], and [31] for more details.

Let $\alpha \colon \Pi \to H_1(M_K) \cong \mathbb{Z}$ be the Abelianization map. For a representation ρ, put $\tilde{\rho} := \mathrm{Ad}\, \rho \otimes \alpha$, that is, $(\mathrm{Ad}\, \rho \otimes \alpha)(x) := \alpha(x)\rho(x) \in \mathrm{SL}(2; \mathbb{C})$ for $x \in \pi_1(M_K)$. Let $\mathscr{T}^K(\tilde{\rho}; t)$ be the Reidemeister torsion of $\tilde{\rho}$, where t is a multiplicative generator of $H_1(M_K)$. Since the chain complex associated with $\tilde{\rho}$ is acyclic if ρ is λ-regular [90, Proposition 3.1.1], the corresponding Reidemeister torsion is well-defined, where λ is the longitude of a knot K. Then the twisted Reidemeister torsion of a λ-regular representation ρ can be calculated by using $\mathscr{T}^K(\tilde{\rho}; t)$.

Theorem 5.2 ([90, Theorem 3.1.2]) *If a representation ρ is λ-regular, then the twisted Reidemeister torsion $\mathbb{T}_\lambda^K(\rho)$ of ρ associated with λ is given as*

$$\mathbb{T}_\lambda^K(\rho) = -\lim_{t \to 1} \frac{\mathcal{T}^K(\tilde{\rho}; t)}{t-1}.$$

Remark 5.1 From [49, Theorem A] $\mathcal{T}^K(\tilde{\rho})$ coincides with the twisted Alexander polynomial of K associated with $\tilde{\rho}$.

From Theorem 5.2 and Remark 5.1, we have the following theorem (see also [48]).

Theorem 5.3 *For a λ-regular representation ρ, we have*

$$\mathbb{T}_\lambda^K(\rho) = \pm \lim_{t \to 1} \frac{\det \tilde{\rho}\left(\frac{\partial r_i}{\partial x_j}\right)}{(t-1)\det \tilde{\rho}(x_l - 1)}.$$

Remark 5.2 We can determine the sign in the formula above. See [21] for details.

If one wants to calculate the twisted Reidemeister torsion associated with the meridian μ, the following formula [73, Théorème 4.1 (ii)] is useful.

Theorem 5.4 ([73, Théorème 4.1 (ii)]) *Suppose that a representation ρ sends the meridian μ to $\begin{pmatrix} e^{u/2} & * \\ 0 & e^{-u/2} \end{pmatrix}$ and the longitude λ to $\begin{pmatrix} e^{v(u)/2} & * \\ 0 & e^{-v(u)/2} \end{pmatrix}$ with u a complex parameter. Then we have*

$$\mathbb{T}_\mu^K(\rho) = \pm \frac{\mathbb{T}_\lambda^K(\rho)}{d\,v(u)/d\,u}.$$

Example 5.9 (Figure-eight knot) We calculate $\mathbb{T}_\mu^{\mathcal{E}}(\rho_{e^{u/2},\pm})$, where $\mathcal{E}$ is the figure-eight knot. Note that in [73, p. 113] J. Porti calculated it in a sophisticated way. See also [20, § 6.3].

From Example 5.1, we have $\pi_1(S^3 \setminus \mathcal{E}) = \langle x, y \mid xy^{-1}x^{-1}yx = yxy^{-1}x^{-1}y \rangle$. In this case there is only one relation $r := xy^{-1}x^{-1}yxy^{-1}xyx^{-1}y^{-1}$ and we have

$$\frac{\partial r}{\partial x} = 1 - xy^{-1}x^{-1} + xy^{-1}x^{-1}y + xy^{-1}x^{-1}yxy^{-1} - xy^{-1}x^{-1}yxy^{-1}xyx^{-1},$$

$$\frac{\partial r}{\partial y} = -xy^{-1} + xy^{-1}x^{-1} - xy^{-1}x^{-1}yxy^{-1} + xy^{-1}x^{-1}yxy^{-1}x$$

$$- xy^{-1}x^{-1}yxy^{-1}xyx^{-1}y^{-1}.$$

The adjoint actions of $\rho_{e^{u/2},\pm}(x)$ are given as follows:

$$\text{Ad}\,\rho_{e^{u/2},\pm}(x)(E) := \rho_{e^{u/2},\pm(x)}^{-1} \cdot E \cdot \rho_{e^{u/2},\pm} = \begin{pmatrix} 0 & e^{-u} \\ 0 & 0 \end{pmatrix},$$

$$\mathrm{Ad}\,\rho_{e^u/2,\pm}(x)(H) = \begin{pmatrix} 1 & 2e^{-u/2} \\ 0 & -1 \end{pmatrix},$$

$$\mathrm{Ad}\,\rho_{e^u/2,\pm}(x)(F) = \begin{pmatrix} -e^{u/2} & -1 \\ e^u & e^{u/2} \end{pmatrix}.$$

So with respect to the basis $\{E, H, F\}$, $\mathrm{Ad}\,\rho_{e^u/2,\pm}(x)$ is given by the 3×3 matrix

$$X := \begin{pmatrix} e^{-u} & 2e^{-u/2} & -1 \\ 0 & 1 & -e^{u/2} \\ 0 & 0 & e^u \end{pmatrix}.$$

Similarly, $\mathrm{Ad}\,\rho_{e^u/2,\pm}(y)$ is given by

$$Y := \begin{pmatrix} e^{-u} & 0 & 0 \\ -e^{-u/2}d_\pm & 1 & 0 \\ -d_\pm^2 & 2e^{u/2}d_\pm & e^u \end{pmatrix}$$

with respect to the same basis.
We also have

$$\mathrm{Ad}\,\rho_{e^u/2,\pm}\left(\frac{\partial r}{\partial x}\right)$$
$$= I_3 - X^{-1}Y^{-1}X + YX^{-1}Y^{-1}X + Y^{-1}XYX^{-1}Y^{-1}X$$
$$- X^{-1}YXY^{-1}XYX^{-1}Y^{-1}X,$$

where I_3 the 3×3 identity matrix. Note that we need to reverse the order of the multiplication.

Put $\tilde{\rho}_{e^u/2,\pm} := \mathrm{Ad}\,\rho_{e^u/2,\pm} \otimes \alpha$. Since $\tilde{\rho}_{e^u/2,\pm}(x) = tX$ and $\tilde{\rho}_{e^u/2,\pm}(y) = tY$, we have

$$\tilde{\rho}_{e^u/2,\pm}\left(\frac{\partial r}{\partial x}\right)$$
$$= I_3 - t^{-1}X^{-1}Y^{-1}X + YX^{-1}Y^{-1}X + Y^{-1}XYX^{-1}Y^{-1}X$$
$$- tX^{-1}YXY^{-1}XYX^{-1}Y^{-1}X$$

and Mathematica tells us

$$\det \tilde{\rho}_{e^u/2,\pm}\left(\frac{\partial r}{\partial x}\right)$$
$$= -t^{-3}e^{-2u}(t-1)^2(t-e^u)(te^u-1)\big(e^u+t(-2+(t-1)e^u-2e^{2u})\big).$$

Since $\det \tilde{\rho}_{e^u/2,\pm}(y-1) = (te^{-u}-1)(t-1)(te^u-1)$, we have

$$\mathbb{T}_\lambda^{\mathscr{E}}\left(\rho_{e^u/2,\pm}\right) = \lim_{t\to 1} \frac{\det \tilde{\rho}_{e^u/2,\pm}\left(\frac{\partial r}{\partial x}\right)}{(t-1)\det \tilde{\rho}_{e^u/2,\pm}(y-1)} = 4\cosh u - 1$$

from Theorem 5.3.

Now we apply Theorem 5.4. From (5.2), we have

$$\frac{d\,v(u)}{d\,u} = \pm 2\frac{d}{d\,u}\log\left(\cosh(2u) - \cosh u - 1 - \sinh u\sqrt{(2\cosh u + 1)(2\cosh u - 3)}\right)$$

$$= \pm\frac{2(1 - 4\cosh u)}{\sqrt{(2\cosh u + 1)(2\cosh u - 3)}}.$$

Therefore we finally have

$$\mathbb{T}_\mu^{\mathscr{E}}\left(\rho_{e^u/2,\pm}\right) = \frac{\mathbb{T}_\lambda^{\mathscr{E}}\left(\rho_{e^u/2,\pm}\right)}{d\,v(u)/d\,u} = \frac{\sqrt{(2\cosh u + 1)(2\cosh u - 3)}}{2}$$

up to a sign.

Example 5.10 (Torus knot) We calculate $\mathbb{T}_\mu^{T(2,2a+1)}(\rho_{e^u/2,j})$.

It is known that any irreducible representation of $\pi_1\left(S^3 \setminus T(p,q)\right)$ is λ-regular and μ-regular (see [20, Example 1]). So the representation $\rho_{e^u/2,j}$ given in Example 5.4 is λ-regular and μ-regular unless $u = \frac{(2k+1)\pi\sqrt{-1}}{2a+1}$.

Putting $r := (xy)^a x (xy)^{-a} y^{-1}$, we have

$$\frac{\partial r}{\partial x} = \sum_{i=0}^{a-1}(xy)^i + (xy)^a\left(1 - x(xy)^{-a}\left(\sum_{i=0}^{a-1}(xy)^i\right)\right).$$

For $z \in \pi_1\left(S^3 \setminus T(2, 2a+1)\right)$, put $\tilde{\rho}_{e^u/2,j}(z) := \alpha(z)\,\mathrm{Ad}\,\rho_{e^u/2,j}(z)$. We also put $X := \tilde{\rho}_{e^u/2,j}(x)$ and $Y := \tilde{\rho}_{e^u/2,j}(y)$. Then we have

$$X = t\begin{pmatrix} e^{-u} & 2e^{-u/2} & -1 \\ 0 & 1 & -e^{u/2} \\ 0 & 0 & e^u \end{pmatrix},$$

$$Y = t\begin{pmatrix} e^{-u} & 0 & 0 \\ e^{-u/2}\left(2\cos\left(\frac{(2j+1)\pi}{2a+1}\right) - 2\cosh u\right) & 1 & 0 \\ -\left(2\cos\left(\frac{(2j+1)\pi}{2a+1}\right) - 2\cosh u\right)^2 & -2e^{u/2}\left(2\cos\left(\frac{(2j+1)\pi}{2a+1}\right) - 2\cosh u\right) & e^u \end{pmatrix},$$

and

$$\tilde{\rho}_{e^{u/2},j}\left(\frac{\partial r}{\partial x}\right) = \sum_{i=0}^{a-1}(YX)^i + \left(I_3 - \left(\sum_{i=0}^{a-1}(YX)^i\right)(YX)^{-a}X\right)(YX)^a.$$

By Mathematica we have

$$\det \tilde{\rho}_{e^{u/2},j}\left(\frac{\partial r}{\partial x}\right) = \frac{(t^{2a+1} - 1)^2(t^{2a+1} + 1)(te^u - 1)(te^{-u} - 1)}{(t + 1)(t^2 - \omega^2)(t^2 - \omega^{-2})}$$

with $\omega := \exp\left(\frac{(2j+1)\pi\sqrt{-1}}{2a+1}\right)$. Since

$$\det \tilde{\rho}_{e^{u/2},j}(y - 1) = (t - 1)(te^u - 1)(te^{-u} - 1),$$

we have

$$\mathbb{T}_\lambda^{T(2,2a+1)}\left(\rho_{e^{u/2},j}\right) = \pm \lim_{t\to1} \frac{\det \tilde{\rho}_{e^{u/2},j}\left(\frac{\partial r}{\partial x}\right)}{(t - 1)\det \rho_{e^{u/2},j}(y - 1)} = \pm\left(\frac{2a + 1}{2\sin\left(\frac{(2j+1)\pi}{2a+1}\right)}\right)^2.$$

Since $d\,v(u)/d\,u = -2(2a + 1)$ from (5.8), we have

$$\mathbb{T}_\mu^{T(2,2a+1)}\left(\rho_{e^{u/2},j}\right) = \frac{\mathbb{T}_\lambda^{T(2,2a+1)}\left(\rho_{e^{u/2},j}\right)}{d\,v(u)/d\,u} = \frac{2a + 1}{8\sin^2\left(\frac{(2j+1)\pi}{2a+1}\right)} \tag{5.23}$$

up to a sign.

Chapter 6
Generalizations of the Volume Conjecture

Abstract In this chapter we show various generalizations of the volume conjecture. Firstly, we introduce the complexification of the conjecture by studying the imaginary part of $\log J_N(K; \exp(2\pi\sqrt{-1}/N))$. We expect the $(SL(2; \mathbb{C}))$ Chern–Simons invariant to appear. Secondly, we refine the conjecture by considering more precise approximation of the colored Jones polynomial. We conjecture that the Reidemeister torsion would appear. Lastly, we perturb $2\pi\sqrt{-1}$ in $\exp(2\pi\sqrt{-1}/N)$ slightly and see what happens to the asymptotic expansion of the colored Jones polynomial. The corresponding topological phenomenon is to perturb the hyperbolic structure of the knot complement, provided the knot is hyperbolic. If the knot is non-hyperbolic we expect various representations of the fundamental group of the knot complement to $SL(2; \mathbb{C})$.

6.1 Complexification

What happens if we remove the absolute value sign in the left hand side of (3.1)? J. Murakami, M. Okamoto, T. Takata, and the authors proposed the following generalization of the Volume Conjecture in [67] and checked it for several knots by using computer.

Conjecture 6.1 (Complexification of the Volume Conjecture) For any knot K, we would have

$$2\pi \lim_{N\to\infty} \frac{\log J_N(K; \exp(2\pi\sqrt{-1}/N))}{N}$$
$$= \mathrm{Vol}(S^3 \setminus K) + 2\pi^2\sqrt{-1}\, \mathrm{cs}(S^3 \setminus K) \pmod{\pi^2\sqrt{-1}\mathbb{Z}}.$$

Here cs is the $(SO(3))$ Chern–Simons invariant associated with the Levi–Civita connection (see (4.4)) when K is hyperbolic. In general the Chern–Simons invariant is defined for a representation of the fundamental group of a three-manifold to

© The Author(s), under exclusive licence to Springer Nature Singapore Pte Ltd. 2018
H. Murakami, Y. Yokota, *Volume Conjecture for Knots*, SpringerBriefs in
Mathematical Physics 30, https://doi.org/10.1007/978-981-13-1150-5_6

$SL(2; \mathbb{C})$ and here we also conjecture that there would exist a canonical representation (the holonomy representation for a hyperbolic knot) even if K is not hyperbolic.
We can also think that cs is defined by the left hand side.

It would be interesting to use an argument similar to Chap. 4 following [69] and
[97] to get a relation to the Chern–Simons invariant.

6.2   Refinement

In this section we try to get a better approximation of $J_N(K; \exp(2\pi\sqrt{-1}/N))$ for
large N. We start with the figure-eight knot $\mathscr{E}$.

6.2.1   Figure-Eight Knot

In this subsection we follow [2] to give a rough sketch.

To do this, we use the quantum dilogarithm $\psi_N(z)$ defined in Chap. 4 (see also
[2, 23, 39]). Putting $z := k/N$ and $1 - k/N$ we have

$$1 - e^{2k\pi\sqrt{-1}/N} = \frac{\psi_N(k/N - 1/(2N))}{\psi_N(k/N + 1/(2N))}$$

and

$$1 - e^{-2k\pi\sqrt{-1}/N} = \frac{\psi_N(1 - k/N - 1/(2N))}{\psi_N(1 - k/N + 1/(2N))}$$

from (4.1). Therefore we have

$$\prod_{k=1}^{j} \left(1 - e^{2k\pi\sqrt{-1}/N}\right)\left(1 - e^{-2k\pi\sqrt{-1}/N}\right)$$

$$= \frac{\psi_N(1/(2N))}{\psi_N((2j+1)/(2N))} \frac{\psi_N(1 - (2j+1)/(2N))}{\psi_N(1 - 1/(2N))}$$

and so we have

$$J_N(\mathscr{E}; \exp(2\pi\sqrt{-1}/N)) = \frac{\psi_N(1/(2N))}{\psi_N(1 - 1/(2N))} \sum_{j=0}^{N-1} \frac{\psi_N(1 - (2j+1)/(2N))}{\psi_N((2j+1)/(2N))}.$$

Since $\dfrac{\psi_N(1/(2N))}{\psi_N(1 - 1/(2N))} = N$ from (4.2), we have

$$\frac{1}{N} J_N(\mathscr{E}; \exp(2\pi\sqrt{-1}/N)) = \sum_{j=0}^{N-1} g_N\left(\frac{2j+1}{2N}\right),$$

where we define

$$g_N(z) := \frac{\psi_N(1-z)}{\psi_N(z)}. \tag{6.1}$$

Note that $g_N(z)$ is defined for z with $-\frac{1}{2N} < \mathrm{Re}\, z < 1 + \frac{1}{2N}$.

For $0 < \varepsilon < 1/(4N)$, let C_ε be the rectangle in the complex plane with vertices $\varepsilon + \sqrt{-1}, \varepsilon - \sqrt{-1}, 1 - \varepsilon - \sqrt{-1}$, and $1 - \varepsilon + \sqrt{-1}$ with counterclockwise orientation. Let C_ε^+ (C_ε^-, respectively) be the polygonal line from $1 - \varepsilon$ to ε (from ε to $1 - \varepsilon$, respectively) along C_ε. Note that the domain of $g_N(w)$ contains C_ε.

Since the set of the poles of $\tan(N\pi z)$ inside C_ε is $\{(2j + 1)/(2N) \mid j = 0, 1, 2, \dots, N - 1\}$ and the residue of each pole is $-1/(N\pi)$, we have

$$\frac{1}{N} J_N(\mathscr{E}; \exp(2\pi\sqrt{-1}/N)) = \frac{\sqrt{-1}N}{2} \int_{C_\varepsilon} \tan(N\pi z) g_N(z)\, dz$$

from the residue theorem. Using the fact that $\tan(w)$ is close to $\sqrt{-1}$ ($-\sqrt{-1}$, respectively) when $\mathrm{Im}(w) \gg 0$ ($-|\mathrm{Im}(w)| \gg 0$, respectively),[1] we can approximate $\int_{C_\varepsilon^\pm} \tan(N\pi z) g_N(z)\, dz$ by $\pm\sqrt{-1} \int_{C_\varepsilon^\pm} g_N(z)\, dz$. In fact we can show that

$$\int_{C_\varepsilon^\pm} \tan(N\pi z) g_N(z)\, dz = \pm\sqrt{-1} \int_{C_\varepsilon^\pm} g_N(z)\, dz + O(N^{-1}).$$

See [2, Equation (4.7)].

Since $g_N(z)$ is analytic inside C_ε we have

$$\frac{1}{N^2} J_N(\mathscr{E}; \exp(2\pi\sqrt{-1}/N)) = \int_P g_N(z)\, dz + O(N^{-1}), \tag{6.2}$$

where P is any path connecting ε and $1 - \varepsilon$ inside the rectangle C_ε.

---

[1] The first author learned it from Kashaev. Writing $\mathrm{Re}\, w = x$ and $\mathrm{Im}\, w = y$, $\tan w = \frac{\cosh y \sin x + \sqrt{-1}\sinh y \cos x}{\cosh y \cos x - \sqrt{-1}\sinh y \sin x}$. So $\tan w \to \sqrt{-1}$ when $y \to \infty$ and $\tan w \to -\sqrt{-1}$ when $y \to -\infty$.

Next we approximate this integral. From (4.3) we see that $g_N(z)$ converges to

$$\exp\left(\frac{N}{2\pi\sqrt{-1}}(\mathscr{L}(1-z) - \mathscr{L}(z))\right).$$

In fact, we can prove the following (see [2, Equation (4.9)]], [64, Proposition 3.2], and (6.5) below).

$$\int_P g_N(z)\,dz = \int_P e^{N\Phi(z)}\,dz + O\left(\frac{\log(N)}{N} \times e^{M_P \times N}\right),\tag{6.3}$$

where

$$\Phi(z) := \frac{1}{2\pi\sqrt{-1}}\left(\mathrm{Li}_2\left(e^{-2\pi\sqrt{-1}z}\right) - \mathrm{Li}_2\left(e^{2\pi\sqrt{-1}z}\right)\right)$$

and M_P is the maximum of $\mathrm{Re}\,\Phi(z)$ along the path P.

Since Li_2 is analytic in the region $\mathbb{C}\setminus(1,\infty)$, the function Φ is analytic in the region $\{0 < \mathrm{Re}\,z < 1\}$.

We will study the asymptotic behavior of $\displaystyle\int_P e^{N\Phi(z)}\,dz$ for large N. To do that we apply the following version of the *saddle point method* due to Ohtsuki.

Proposition 6.1 ([71, Proposition 3.2]) *Let $\Upsilon(w)$ be a holomorphic function of the form $\Upsilon(w) = aw^2 + \sum_{i \geq 3} b_i w^i$ in a neighborhood of 0 with $a \neq 0$. The domain $\{w \in \mathbb{C} \mid \mathrm{Re}\,\Upsilon(w) < 0\}$ has two connected components D_0 and D_1 in a neighborhood of 0. Choose $p_0 \in D_0$ and $p_1 \in D_1$ and let C be a path from p_0 to p_1.*

Then we have

$$\int_C e^{N\Upsilon(w)}\,dw = \frac{\sqrt{\pi}}{\sqrt{-a}\sqrt{N}} + O(N^{-1}),$$

where we choose $\sqrt{-a}$ so that $\mathrm{Re}(p_1\sqrt{-a}) > 0$.

Since we have

$$\frac{d\,\Phi(z)}{d\,z} = \log(2 - e^{2\pi\sqrt{-1}z} - e^{-2\pi\sqrt{-1}z}),$$

and

$$\frac{d^2\,\Phi(z)}{d\,z^2} = \frac{2\pi\sqrt{-1}(e^{2\pi\sqrt{-1}z} + 1)}{e^{2\pi\sqrt{-1}z} - 1},$$

the function $\Phi(z)$ has the following Taylor expansion around $z = \dfrac{5}{6}$:

$$\Phi(z) = \Phi\left(\frac{5}{6}\right) - \sqrt{3}\pi(z - \frac{5}{6})^2 + \cdots.$$

Note that

$$\Phi(5/6) = \frac{-2}{\pi}\Lambda(5\pi/6) = \frac{\mathrm{Vol}(S^3 \setminus \mathscr{E})}{2\pi} = 0.323\cdots > 0 \qquad (6.4)$$

from the following identities (see for example [60]):

$$\mathrm{Li}_2(z^{-1}) = -\mathrm{Li}_2(z) - \frac{\pi^2}{6} - \frac{1}{2}(\log(-z))^2, \qquad (6.5)$$

$$\mathrm{Li}_2(e^{2\sqrt{-1}\theta}) = \frac{\pi^2}{6} - \theta(\pi - \theta) + 2\sqrt{-1}\Lambda(\theta) \quad (0 \le \theta \le \pi) \qquad (6.6)$$

and (3.4).

If we put $\Upsilon(w) := \Phi\left(w + \frac{5}{6}\right) - \Phi\left(\frac{5}{6}\right)$, then $\Upsilon(w)$ satisfies the condition of Proposition 6.1 with $a = -\sqrt{3}\pi$. See Figs. 6.1 and 6.2. We choose C as the line segment $[\varepsilon, 1-\varepsilon]$, and put $p_0 := \varepsilon - \frac{5}{6}$ and $p_1 := \frac{1}{6} - \varepsilon$. Then from Proposition 6.1, we have

$$\int_C e^{N\Upsilon(w)}\, dw = \frac{1}{3^{1/4}\sqrt{N}} + O(N^{-1}).$$

Since we have

$$\int_\varepsilon^{1-\varepsilon} e^{N\Phi(z)}\, dz = e^{\Phi(5/6)N}\int_C e^{N\Upsilon(w)}\, dw,$$

we have

$$\int_\varepsilon^{1-\varepsilon} g_N(z)\, dz = \frac{e^{\Phi(5/6)N}}{3^{1/4}\sqrt{N}} + O\left(\frac{\log N}{N} \times e^{(\max_{\varepsilon \le z \le 1-\varepsilon}\, \mathrm{Re}\,\Phi(z)) \times N}\right)$$

$$= \frac{e^{\Phi(5/6)N}}{3^{1/4}\sqrt{N}} + O\left(\frac{\log N}{N} \times e^{\Phi(5/6)N}\right)$$

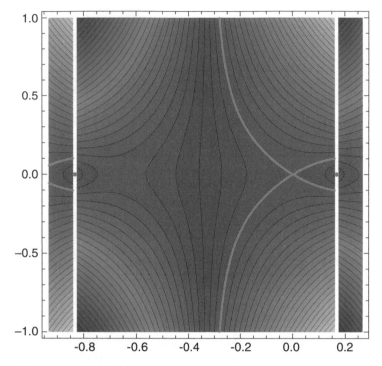

Fig. 6.1 Contour graph of $\operatorname{Re}\Upsilon(w)$. The blue lines indicate the set $\{w \in \mathbb{C} \mid \operatorname{Re}\Upsilon(w) = 0\}$ (Plotted by Mathematica)

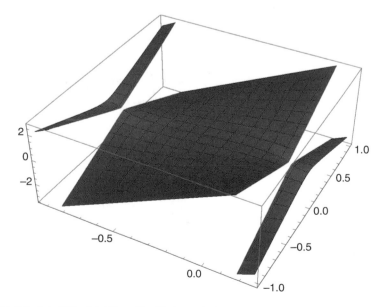

Fig. 6.2 3D plot of Fig. 6.1 (Plotted by Mathematica)

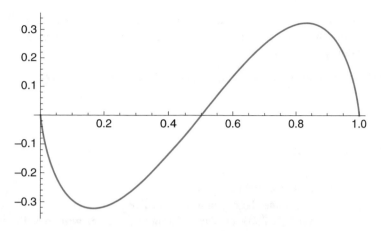

Fig. 6.3 Graph of Re $\Phi(z)$ plotted by Mathematica. It takes the maximum at $z = 5/6$

from (6.3). See Fig. 6.3. Therefore from (6.2) and (6.4) we have

$$J_N(\mathscr{E}; \exp(2\pi\sqrt{-1}/N))$$

$$= 3^{-1/4} N^{3/2} \exp\left(\mathrm{Vol}(S^3 \setminus \mathscr{E}) \frac{N}{2\pi}\right) + O\left(N \log N \times e^{\mathrm{Vol}(S^3\setminus\mathscr{E})\frac{N}{2\pi}}\right),$$

which is due to J. Andersen and S. Hansen [2, Lemma 4]. Putting $T_0 := \sqrt{\frac{2}{\sqrt{-3}}}$ and $S_0 := \sqrt{-1}\,\mathrm{Vol}(S^3 \setminus \mathscr{E})$ we can re-write the formula above as follows:

$$J_N(\mathscr{E}; \exp(2\pi\sqrt{-1}/N))$$

$$= 2\pi^{3/2} T_0 \left(\frac{N}{2\pi\sqrt{-1}}\right)^{3/2} \exp\left(\frac{N}{2\pi\sqrt{-1}} \times S_0\right) + O\left(N \log N \times e^{\mathrm{Vol}(S^3\setminus\mathscr{E})\frac{N}{2\pi}}\right).$$

From Example 5.6 we know that $\sqrt{-1}S_0$ is the Chern–Simons invariant of the holonomy representation. We also know from Example 5.9 that $(T_0)^{-2}$ is the twisted Reidemeister torsion of the holonomy representation associated with the meridian, up to a sign.

We expect a similar result holds for any hyperbolic knot [29, 71].

Conjecture 6.2 (Refinement of the Volume Conjecture for hyperbolic knots) Let K be a hyperbolic knot. Then we have the following asymptotic equivalence as $N \to \infty$:

$$J_N(K; \exp(2\pi\sqrt{-1}/N))$$

$$\underset{N\to\infty}{\sim} 2\pi^{3/2} T_0 \left(\frac{N}{2\pi\sqrt{-1}}\right)^{3/2} \exp\left(\frac{N}{2\pi\sqrt{-1}} \times S_0\right),$$

where $(T_0)^{-2}$ is the twisted Reidemeister torsion of the holonomy representation associated with the meridian, and $\sqrt{-1}S_0$ is the $SL(2; \mathbb{C})$ Chern–Simons invariant of the holonomy representation.

So far the conjecture is proved for the hyperbolic knots 4_1 [2], 5_2 [71], 6_1, 6_2, and 6_3 [72].

6.2.2   Torus Knot

Next we consider the torus knot of type $(2, 2a + 1)$.

In Sect. 3.3, we calculate the asymptotic behavior of the colored Jones polynomial of the torus knot $T(2, 2a + 1)$. The formula (3.7) can be written as follows.

$$
J_N\left(T(2, 2a + 1); e^{2\pi\sqrt{-1}/N}\right)
$$

$$
= \frac{\pi^{3/2}}{4(2a + 1)} \left(\frac{N}{2\pi\sqrt{-1}}\right)^{3/2} \left(\sum_{k=0}^{2a}(-1)^k(2k + 1)^2 T_k \exp\left(\frac{N}{2\pi\sqrt{-1}} \times S_k\right)\right)
$$

$$
+ O(N^{1/2}),
$$

where

$$
T_k = \frac{2\sqrt{2}\sin\left(\frac{(2k+1)\pi}{2a+1}\right)}{\sqrt{2a + 1}},
$$

and

$$
S_k = \frac{((2k + 1) - 2(2a + 1))^2\pi^2}{2(2a + 1)}.
$$

Note that S_k coincides with the Chern–Simons invariant of the representation $\rho_{0,k}$ modulo $\pi^2\mathbb{Z}$ and that T_k^{-2} is the twisted Reidemeister torsion of $\rho_{0,k}$ associated with the meridian up to sign.

6.3   Parametrization

In Sect. 3.3 we calculate the asymptotic behavior of $J_N(T(2, 2a+1); e^{2\pi\sqrt{-1}/N})$ but the reader may find that it would be easier to calculate it in the case where $2\pi\sqrt{-1}$ is replaced with ξ for generic ξ. In this section we consider the asymptotic behavior of $J_N(K; e^{\xi/N})$ with $\xi \neq 2\pi\sqrt{-1}$. We start with the torus knot of type $(2, 2a + 1)$.

6.3.1 Torus Knot

In (3.5) we calculate

$$(e^{\xi/2} - e^{-\xi/2}) J_N(T(2, 2a+1); e^{\xi/N})$$

$$= \exp\left(\frac{-(2a+1)(N^2-1)\xi}{2N}\right) \exp\left(-\frac{\xi}{4N}\left(\frac{2a+1}{2} + \frac{2}{2a+1}\right)\right) \sqrt{\frac{N}{2(2a+1)\xi\pi}}$$

$$\times \int_{C_\varphi} \frac{\sinh\left(\frac{x}{2a+1}\right)}{\cosh\left(\frac{x}{2}\right)} \exp\left(\frac{-N}{2(2a+1)\xi}x^2 + Nx\right) dx$$

for any $\xi \in \mathbb{C}$. We can apply the saddle point method (Theorem 3.1) again to obtain the asymptotic behavior of the formula above, which is easier than the case where $\xi = 2\pi\sqrt{-1}$. The integral becomes

$$e^{(2a+1)\xi N/2} \int_{C_\varphi} \frac{\sinh\left(\frac{x}{2a+1}\right)}{\cosh\left(\frac{x}{2}\right)} \exp\left(\frac{-N}{2(2a+1)\xi}(x - (2a+1)\xi)^2\right) dx.$$

Put $\tilde{C}_\varphi := \{t \exp(\varphi\sqrt{-1}) + (2a+1)\xi \mid t \in \mathbb{R}\}$. By the residue theorem we have

$$\int_{C_\varphi} \frac{\sinh\left(\frac{x}{2a+1}\right)}{\cosh\left(\frac{x}{2}\right)} \exp\left(\frac{-N}{2(2a+1)\xi}(x - (2a+1)\xi)^2\right) dx$$

$$= \int_{\tilde{C}_\varphi} \frac{\sinh\left(\frac{x}{2a+1}\right)}{\cosh\left(\frac{x}{2}\right)} \exp\left(\frac{-N}{2(2a+1)\xi}(x - (2a+1)\xi)^2\right) dx$$

$$+ 2\pi\sqrt{-1} \sum_k \mathrm{Res}\left(\frac{\sinh\left(\frac{x}{2a+1}\right)}{\cosh\left(\frac{x}{2}\right)} \exp\left(\frac{-N}{2(2a+1)\xi}(x-(2a+1)\xi)^2\right); x = (2k+1)\pi\sqrt{-1}\right)$$

$$= \int_{\tilde{C}_\varphi} \frac{\sinh\left(\frac{x}{2a+1}\right)}{\cosh\left(\frac{x}{2}\right)} \exp\left(\frac{-N}{2(2a+1)\xi}(x - (2a+1)\xi)^2\right) dx$$

$$+ 2\pi\sqrt{-1} \sum_k (-1)^k 2\sin\left(\frac{(2k+1)\pi}{2a+1}\right) \exp\left(\frac{-N}{2(2a+1)\xi}((2k+1)\pi\sqrt{-1} - (2a+1)\xi)^2\right)$$

(Putting $y := x - (2a + 1)\xi$)

$$= \int_{C_\varphi} \frac{\sinh\left(\frac{y+(2a+1)\xi}{2a+1}\right)}{\cosh\left(\frac{y+(2a+1)\xi}{2}\right)} \exp\left(\frac{-N}{2(2a+1)\xi}y^2\right) dx$$

$$+ 2\pi\sqrt{-1}\sum_k(-1)^k 2\sin\left(\frac{(2k+1)\pi}{2a+1}\right)\exp\left(\frac{-N}{2(2a+1)\xi}((2k+1)\pi\sqrt{-1}-(2a+1)\xi)^2\right)$$

$$= \sqrt{\frac{2(2a+1)\xi\pi}{N}} \frac{\sinh\xi}{\cosh\left(\frac{(2a+1)\xi}{2}\right)}$$

$$+ 2\pi\sqrt{-1}\sum_k(-1)^k 2\sin\left(\frac{(2k+1)\pi}{2a+1}\right)\exp\left(\frac{-N}{2(2a+1)\xi}((2k+1)\pi\sqrt{-1}-(2a+1)\xi)^2\right)$$

$$+ O(N^{-1}),$$

where we use Theorem 3.1 in the last equality. Here the summation is over k such that $(2k+1)\pi\sqrt{-1}$ is between C_φ and $\tilde{C}_\varphi$. Therefore we have

$$2\sinh(\xi/2)\exp\left(\frac{\xi}{4N}\left(\frac{2a+1}{2}+\frac{2}{2a+1}-2(2a+1)\right)\right) J_N(T(2,2a+1); e^{\xi/N})$$

$$= \frac{\sinh\xi}{\cosh\left(\frac{(2a+1)\xi}{2}\right)}$$

$$+ 2\pi\sqrt{-1}\sqrt{\frac{N}{2(2a+1)\xi\pi}}$$

$$\times \sum_k(-1)^k 2\sin\left(\frac{(2k+1)\pi}{2a+1}\right)\exp\left(\frac{-N}{2(2a+1)\xi}((2k+1)\pi\sqrt{-1}-(2a+1)\xi)^2\right)$$

$$+ O(N^{-1/2})$$

$$= \frac{-2\sinh(u/2)}{\Delta(T(2,2a+1); \exp u)}+\sqrt{-\pi}\sum_k(-1)^k T_k\left(\frac{N}{2\pi\sqrt{-1}+u}\right)^{1/2}\exp\left(\frac{N}{2\pi\sqrt{-1}+u}S_k(u)\right)$$

$$+ O(N^{-1/2}),$$

where

$$S_k(u) := \frac{-((2k+1)\pi\sqrt{-1}-(2a+1)(2\pi\sqrt{-1}+u))^2}{2(2a+1)}$$

$$T_k := \frac{2\sqrt{2}\sin\left(\frac{(2k+1)\pi}{2a+1}\right)}{\sqrt{2a+1}}.$$

Observe that

- T_k^{-2} coincides with the twisted Reidemeister torsion $\mathbb{T}_\mu^{T(2,2a+1)}(\rho_{e^{u/2},k})$ with $\xi = 2\pi\sqrt{-1} + u$ up to sign.
- $S_k(u) - \pi\sqrt{-1}u - \dfrac{1}{4}uv_k(u)$ coincides with the Chern–Simons invariant of $\rho_{e^{u/2},k}$
  with $v_k(u) := 2\dfrac{d\,\tilde{S}_k(u)}{d\,u}\Big|_{\xi:=2\pi\sqrt{-1}+u} - 2\pi\sqrt{-1}$ when we put $l := k - 2a - 1$
  and $j := k$ in (5.21).
- $\left(\dfrac{-2\sinh(u/2)}{\Delta(T(2,2a+1);\exp u)}\right)^{-2}$ is the twisted Reidemeister torsion of the Abelian
  representation $\rho_{e^{u/2}}^{A}$ (Proposition 5.1).

Similar results are known for $(2, 2b+1)$ cable of the torus knot of type $T(2, 2a+1)$ [65].

6.3.2 Figure-Eight Knot

As in the case of the torus knots, we now change $\exp(2\pi\sqrt{-1}/N)$ to $\exp((2\pi\sqrt{-1} + u)/N)$ for a complex parameter u in the case of the figure-eight knot. We expect in the large N asymptotic, the Chern–Simons invariant and the twisted Reidemeister torsion appear.

As in Chap. 4, we define

$$\psi_{N,u}(z) := \exp\left(\frac{1}{4}\int_{-\infty}^{\infty}\frac{e^{(2z-1)t}}{t\sinh(t)\sinh(\gamma t)}\,dt\right)$$

for a real number u with $0 < u < \log((3 + \sqrt{5})/2)$, where we put $\gamma := \dfrac{2\pi - \sqrt{-1}u}{2\pi N}$. The integral is defined for z with $|\mathrm{Re}(2z - 1)| < 1 + 1/N$. Note that $\psi_N(z) = \psi_{N,0}(z)$.

We have the following formula as (4.1):

$$\frac{\psi_{N,u}(z - \gamma/2)}{\psi_{N,u}(z + \gamma/2)} = 1 - e^{2\pi\sqrt{-1}z}.$$

Using this formula we have

$$J_N(\mathscr{E}; \exp(\xi/N))$$

$$= \frac{\psi_{N,u}(\gamma/2 - \sqrt{-1}u/(2\pi))}{\psi_{N,u}(1 - \gamma/2 - \sqrt{-1}u/(2\pi))}\sum_{k=0}^{N-1}e^{-ku}\frac{\psi_{N,u}(1 - \sqrt{-1}u/(2\pi) - (2k+1)\gamma/2)}{\psi_{N,u}(-\sqrt{-1}u/(2\pi) + (2k+1)\gamma/2)}.$$

Since we can prove (see [64][2])

$$\frac{\psi_{N,u}(\gamma/2 - \sqrt{-1}u/(2\pi))}{\psi_{N,u}(1 - \gamma/2 - \sqrt{-1}u/(2\pi))} = \frac{e^{u\pi/\gamma} - 1}{e^u - 1},$$

we have

$$J_N(\mathscr{E}; e^{(2\pi\sqrt{-1}+u)/N}) = \frac{e^{2\pi u\sqrt{-1}N/\xi} - 1}{2\sinh(u/2)} \sum_{k=0}^{N-1} g_{N,u}\left(\frac{2k+1}{2N}\right),$$

where we put

$$g_{N,u}(z) := e^{-Nuz}\frac{\psi_{N,u}(1 - \sqrt{-1}u/(2\pi) + \sqrt{-1}\xi z/(2\pi))}{\psi_{N,u}(-\sqrt{-1}u/(2\pi) - \sqrt{-1}\xi z/(2\pi))},$$

with $\xi := 2\pi\sqrt{-1} + u$. Note that $g_{N,u}(z)$ is defined in the strip $\{z \in \mathbb{C} \mid -\frac{2\pi\,\mathrm{Re}\,z}{u} - \frac{\pi}{Nu} < \mathrm{Im}\,z < -\frac{2\pi\,\mathrm{Re}\,z}{u} + \frac{2\pi}{u} + \frac{\pi}{Nu}\}$.

In a way similar to the case where $u = 0$, we can prove that

$$\frac{2\sinh(u/2)}{N(e^{2\pi u\sqrt{-1}N/\xi} - 1)} J_N(\mathscr{E}; \exp(\xi/N)) = \int_P g_{N,u}(z)\,dz + O(N^{-1})$$

for any path P from ε to $1 - \varepsilon$ with $0 < \varepsilon < \frac{1}{4N}$. Putting

$$\Phi_u(z) := \frac{1}{2\pi\sqrt{-1}+u}\left(\mathrm{Li}_2(e^{u-(2\pi\sqrt{-1}+u)z}) - \mathrm{Li}_2(e^{u+(2\pi\sqrt{-1}+u)z})\right) - uz,$$

we can also prove

$$\int_P g_{N,u}(z)\,dz = \int_P e^{N\Phi_u(z)}\,dz + O\left(\frac{\log N}{N} \times e^{M_P \times N}\right),$$

where M_P is the maximum of $\mathrm{Re}\,\Phi_u(z)$ on P.

Now we would like to apply the saddle point method to the integral. We put

$$z_0(u) := \frac{\varphi(u) + 2\pi\sqrt{-1}}{2\pi\sqrt{-1}+u},$$

where $\varphi(u)$ is defined in (5.17). Note that $z_0(u)$ is in the first quadrant because $\varphi(u)$ is purely imaginary and $-\pi/3 < \mathrm{Im}\,\varphi(u) < 0$ (see Example 5.6).

---

[2]The proof in [64] is wrong but the statement remains true, which was informed by Ka Ho Wong.

Since we have

$$\frac{d\,\Phi_u(z)}{d\,z} = \log(1 - e^{u-(2\pi\sqrt{-1}+u)z}) + \log(1 - e^{u+(2\pi\sqrt{-1}+u)z}) - u,$$

we see that $\frac{d\,\Phi_u(z_0(u))}{d\,z} = 0$. Moreover we have

$$\xi \Phi_u(z_0(u)) = \mathrm{Li}_2(e^{u-\varphi(u)}) - \mathrm{Li}_2(e^{u+\varphi(u)}) - u(\varphi(u) + 2\pi\sqrt{-1})$$

and

$$\frac{d^2\,\Phi_u(z_0(u))}{d\,z^2} = \xi\sqrt{(2\cosh(u) + 1)(2\cosh(u) - 3)}$$

from (5.18). So we can write $\Phi_u(z)$ in the following form:

$$\Phi_u(z) = \frac{S(u) - 2\pi\sqrt{-1}u}{\xi} + \frac{1}{2}\xi\sqrt{(2\cosh(u) + 1)(2\cosh(u) - 3)}(z - z_0(u))^2$$
$$+ O((z - z_0(u))^3),$$

where we put

$$S(u) := \mathrm{Li}_2(e^{u-\varphi(u)}) - \mathrm{Li}_2(e^{u+\varphi(u)}) - u\varphi(u)$$

as (5.16). We can prove that $\mathrm{Re}\,\frac{S(u)-2\pi\sqrt{-1}u}{\xi} > 0$. See [64, Lemma 3.5].

Putting $\Upsilon_u(w) := \Phi_u(w + z_0(u)) - \frac{S(u)-2\pi\sqrt{-1}u}{\xi}$, we have

$$\Upsilon_u(w) = \frac{1}{2}\xi\sqrt{(2\cosh(u) + 1)(2\cosh(u) - 3)}w^2 + O(w^3).$$

Now we can apply Proposition 6.1 to $\Upsilon_u(w)$, $p_0 := \varepsilon - z_0(u)$, $p_1 := 1 - \varepsilon - z_0(u)$. See Figs. 6.4 and 6.5. We choose a path P' so that it starts at p_0, ends at p_1 and it passes though $z_0(u)$. Then we have

$$\int_{P'} e^{N\Upsilon_u(w)}\,dw = \frac{\sqrt{2}}{e^{-\pi\sqrt{-1}/4}\sqrt{\xi}\,((2\cosh u + 1)(3 - 2\cosh u))^{1/4}}\sqrt{\frac{\pi}{N}} + O(N^{-1}),$$

where we choose the square root according to Proposition 6.1. Since $\Phi_u(z) = \Upsilon_u(z - z_0(u)) + (S(u) - 2\pi\sqrt{-1}u)/\xi$, we have

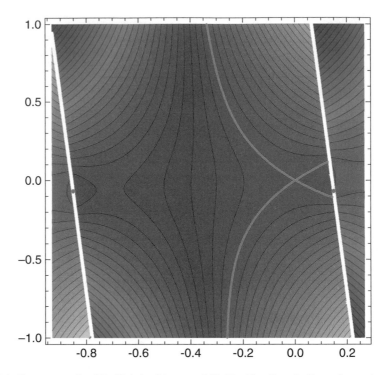

Fig. 6.4 Contour graph of Re $\Upsilon_u(w)$ with $u := 1/2$. The blue lines indicate the set $\{w \in \mathbb{C} \mid \mathrm{Re}\,\Upsilon_u(w) = 0\}$ (Plotted by Mathematica)

$$2\sinh(u/2)\,J_N(\mathscr{E};e^{\xi/N})$$

$$=Ne^{NS(u)/\xi}(1-e^{-2\pi uN/\xi})\int_{P'}e^{N\Upsilon_u(w)}\,dw + O\left(\frac{\log N}{N} \times e^{\frac{S(u)}{\xi}\times N}\right)$$

$$=e^{\pi\sqrt{-1}/4}\sqrt{\pi}\sqrt{\frac{2}{\sqrt{(2\cosh u+1)(3-2\cosh u)}}}\sqrt{\frac{N}{\xi}}e^{NS(u)/\xi}$$

$$+ O\left(\frac{\log N}{N} \times e^{\frac{S(u)}{\xi}\times N}\right),$$

since $\Upsilon_u(w)$ takes its maximum at $z_0(u)$.

Therefore we finally have

$$J_N\left(E;\exp(\xi/N)\right)=\frac{\sqrt{\pi}}{2\sinh(u/2)}T(u)\sqrt{\frac{N}{\xi}}\exp\left(\frac{N}{\xi}S(u)\right)+O\left(\frac{\log N}{N}\times e^{\frac{S(u)}{\xi}\times N}\right),$$

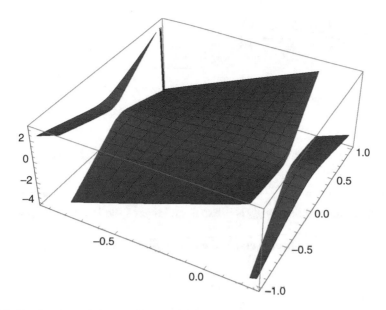

Fig. 6.5 3D plot of Fig. 6.5 (Plotted by Mathematica)

where

$$T(u) := e^{\pi\sqrt{-1}/4}\sqrt{\frac{2}{\sqrt{(2\cosh u + 1)(3 - 2\cosh u)}}}.$$

From Example 5.9, we see that $T(u)^{-2}$ equals the twisted Reidemeister torsion of $\rho_{u,\pm}$ associated with the meridian up to a sign.

From Example 5.6 we see that

- $v(u) = 2\dfrac{d\,S(u)}{d\,u} - 2\pi\sqrt{-1},$
- $S(u) - \pi\sqrt{-1}u - \dfrac{uv(u)}{4}$ equals $\mathrm{CS}_{u,v(u)}([\rho_{e^{u/2}},+]).$

If u is not real, we can prove a similar result but only for the Chern–Simons invariant when $|u|$ is small and not purely imaginary. See [68] for details.

We expect a similar result for any hyperbolic knot K.

Conjecture 6.3 (Parametrization of the Volume Conjecture [19, 29]*)* Suppose that K is hyperbolic. If $|u|$ ($u \neq 0$) is sufficiently small and not purely imaginary, the following asymptotic equivalence holds.

$$J_N\left(K; \exp\left(\frac{2\pi\sqrt{-1}+u}{N}\right)\right)$$

$$\underset{N\to\infty}{\sim} \frac{\sqrt{\pi}}{2\sinh(u/2)} T(u) \sqrt{\frac{N}{2\pi\sqrt{-1}+u}} \exp\left(\frac{N}{2\pi\sqrt{-1}+u} S(u)\right).$$

Here

- $T(u)^{-2}$ is the Reidemeister torsion of an irreducible representation ρ_u associated with the meridian,
- if we put $v(u) := 2\dfrac{d\,S(u)}{d\,u} - 2\pi\sqrt{-1}$, then the ratio of the eigenvalues of the image of the meridian by ρ_u is $\pm\exp u$, and that of the longitude is $\pm\exp v(u)$, and
- $S(u) - \pi\sqrt{-1}u - \frac{1}{4}uv(u)$ equals the Chern–Simons invariant $\mathrm{CS}_{u,v(u)}([\rho_u])$.

6.4   Miscellaneous Results

In this section we describe miscellaneous results about the asymptotic behaviors of the colored Jones polynomials of a knot. We will not give proofs.

First of all we consider the figure-eight knot $\mathcal{E}$. Let θ be a real number and we study the asymptotic behavior of the N-dimensional colored Jones polynomial evaluated at $\exp(\theta/N)$. If $\theta > \operatorname{arccosh}(3/2) = \log\big((3+\sqrt{5})/2\big)$, we can apply a technique similar to Sect. 3.2 to prove the following.

Proposition 6.2 ([62]) *If $\theta > \log\big((3+\sqrt{5})/2\big)$, then we have*

$$\lim_{N\to\infty} \frac{\log J_N(\mathcal{E}; \exp(\theta/N))}{N} = \frac{\tilde{S}(\theta)}{\theta},$$

where we put

$$\tilde{S}(\theta) := \operatorname{Li}_2(e^{-\tilde{\varphi}(\theta)-\theta}) - \operatorname{Li}_2(e^{\tilde{\varphi}(\theta)-\theta}) + \theta\tilde{\varphi}(\theta)$$

with

$$\tilde{\varphi}(\theta) := \operatorname{arccosh}\left(\cosh(\theta) - \frac{1}{2}\right)$$

$$= \log\left(\frac{1}{2}\left(2\cosh(\theta) - 1 + \sqrt{(2\cosh(\theta)+1)(2\cosh(\theta)-3)}\right)\right).$$

Note that we do not need to worry about the branches of the square root, the logarithm, or the dilogarithm (just use the usual ones), since $\theta \geq \operatorname{arccosh}(3/2)$.

When $|\theta| = \text{arccosh}(3/2)$, we can prove that $J_N(\mathscr{E}; \exp(\theta)/N)$ grows polynomially. In fact the following proposition holds.

Proposition 6.3 ([33]) *Let $\Gamma(z)$ be the gamma function. We have*

$$J_N(\mathscr{E}; \exp(\pm\,\text{arccosh}(3/2)/N)) \underset{N\to\infty}{\sim} \frac{\Gamma(1/3)}{\left(3\,\text{arccosh}(3/2)\right)^{2/3}} N^{2/3}.$$

This would correspond to an affine representation. See Definition 5.6 and Example 5.3. In fact for a torus knot, the colored Jones polynomial also has polynomial growth for affine representations [33]. Let $T(p, q)$ be the (p, q)-torus knot, where p and q are coprime and greater than or equal to 2. Then we have

$$J_N\big(T(p,q); \exp(\pm 2\pi\sqrt{-1}/(pqN))\big) \underset{N\to\infty}{\sim} \frac{\sin(\pi/p)\sin(\pi/q)}{\sqrt{2}\sin\big(\pi/(pq)\big)\exp(\pm\pi\sqrt{-1}/4)} N^{1/2}.$$

Note that $\exp(\pm 2\pi\sqrt{-1}/(pq))$ is a zero of the Alexander polynomial of $T(p, q)$. See Example 5.4 for $p = 2$.

When $|\theta| < \text{arccosh}(3/2)$, then $J_N(\mathscr{E}; \exp(\theta/N))$ converges.

Proposition 6.4 ([63]) *If $|\theta| < \text{arccosh}(3/2)$, we have*

$$\lim_{N\to\infty} J_N(\mathscr{E}; \exp(\theta/N)) = \frac{1}{\Delta(\mathscr{E}; \theta)}.$$

In general, the following holds for any knot.

Theorem 6.1 ([26, 63]) *For any knot K there exists an open neighborhood $\mathscr{U}_K \subset \mathbb{C}$ of 0 such that if $a \in \mathscr{U}_K$, then we have*

$$\lim_{N\to\infty} J_N\big(K; \exp(a/N)\big) = \frac{1}{\Delta(K; \exp a)},$$

where $\Delta(K; t)$ is the Alexander polynomial of K, which is normalized so that $\Delta(K; t^{-1}) = \Delta(K; t)$ and $\Delta(K; 1) = 1$.

Theorem 6.1 would correspond to an Abelian representation (Definition 5.4).

6.5 Final Remarks

In this last section we summarize the volume conjecture and its generalizations.

- When K is a hyperbolic knot, then we conjecture that $J_N\big(K; \exp(\xi/N)\big)$

  - grows exponentially when ξ is close to $2\pi\sqrt{-1}$. Moreover the asymptotic expansion determines the $SL(2; \mathbb{C})$ Chern–Simons invariant and the Reide-

meister torsion, both associated with a representation that corresponds to a
hyperbolic structure of the knot complement. In particular, if $\xi = 2\pi\sqrt{-1}$ it
corresponds to the complete hyperbolic structure.
- converges to $1/\Delta(K; \exp\xi)$ when ξ is close to 0, where $\Delta(K; t)$ is the
  Alexander polynomial of the knot K. Moreover ξ corresponds to an Abelian
  representation $\mu \mapsto \begin{pmatrix} \exp(\xi/2) & 0 \\ 0 & \exp(-\xi/2) \end{pmatrix}$, where μ is the meridian, and
- diverges polynomially when $\exp\xi$ is a zero of the Alexander polynomial (and
  ξ is close to 0).

as a function of N.
- When K is not a hyperbolic knot, then we conjecture that the asymptotic
  expansion of $J_N(K; \exp(\xi/N))$

  - splits into summands each of which corresponds to a representation from
    $\pi_1(S^3 \setminus K)$ to SL(2; $\mathbb{C}$). Moreover each summand determines the SL(2; $\mathbb{C}$)
    Chern–Simons invariant and the Reidemeister torsion, both associated with
    the representation.
  - converges to $1/\Delta(K; \exp\xi)$ when ξ is close to 0, where $\Delta(K; t)$ is the
    Alexander polynomial of the knot K. Moreover ξ corresponds to an Abelian
    representation $\mu \mapsto \begin{pmatrix} \exp(\xi/2) & 0 \\ 0 & \exp(-\xi/2) \end{pmatrix}$, where μ is the meridian, and
  - diverges polynomially when $\exp\xi$ is a zero of the Alexander polynomial (and
    ξ is close to 0).

as a function of N.

Remark 6.1 Let M be a closed oriented three-manifold. Due to Witten [88], the
Witten–Reshetikhin–Turaev invariant $Z_k(M)$ of level k is defined as follows:

$$Z_k(M) = \int e^{2\pi(k+2)\sqrt{-1}\, \mathrm{cs}_M(A)} \mathscr{D}A,$$

where the integral is the path integral, that is, "integral over all SU(2) connections",
and $\mathrm{cs}_M(A)$ is the Chern–Simons invariant defined as in (5.9). By the saddle point
method, we can expect that this "integral" is approximated as the sum over all flat
connections:

$$\sum_{\alpha:\text{flat connection}} \text{constant} \times \sqrt{T_\alpha(M)} e^{2\pi\sqrt{-1}(k+2)\, \mathrm{cs}_M(A_\alpha)}$$

for large k, where $T_\alpha(M)$ is the Reidemeister torsion associated with the irreducible
representation induced by the flat connection A_α. See [88, Section 2], [9, Chapter 7]
and [50, Section 3.2] for details.

We will compare this physical idea with our situation. As explained so far we expect that the N-dimensional colored Jones polynomial $J_N\big(K; \exp((2\pi\sqrt{-1} + u)/N)\big)$ can be approximated as

$$\int e^{NF(\mathbf{z})} d\mathbf{z},$$

where $\mathbf{z}$ is a set of parameters. Then by using the saddle point method, this can be further approximated as

$$\sum_{\alpha} (\text{polynomial in } N) \times \sqrt{H(\mathbf{z}_{\alpha})} e^{NF(\mathbf{z}_{\alpha})}$$

for large N, where the $\mathbf{z}_{\alpha}$ are the saddle points.

Using Witten's idea as described above, we expect

- the set $\{\mathbf{z}_{\alpha}\}$ corresponds to the set of irreducible representations from the fundamental group of the knot complement to $\mathrm{SL}(2; \mathbb{C})$,
- $F(\mathbf{z}_{\alpha})$ is (a multiple of) the Chern–Simons invariant associated with the irreducible representation corresponding to $\mathbf{z}_{\alpha}$, and
- $H(\mathbf{z}_{\alpha})$ is the Reidemeister torsion associated with the irreducible representation corresponding to $\mathbf{z}_{\alpha}$.

Remark 6.2 Recently other types of 'volume conjectures' were proposed [3–5, 16].

- A conjecture for a link invariant $J_{M,K}(\hbar, x)$ based on the Teichmüller Topological Quantum Field Theory by Andersen and Kashaev. Here K is a knot in a closed, oriented three-manifold M, $\hbar$ is a positive real number and x is a real number. It says that

$$\lim_{\hbar \to 0} 2\pi \hbar \log |J_{M,K}(\hbar, 0)| = -\mathrm{Vol}(M \setminus K)$$

  if $M \setminus K$ possesses a complete hyperbolic structure. Note that this conjecture is true for $(S^3, 4_1)$ and $(S^3, 5_2)$.

  See also [6, 7] for further developments.
- A conjecture for quantum invariants of three-manifolds by Q. Chen and T. Yang [16]. Here they use roots of unity other than usually used in the Turaev–Viro invariants and Witten–Reshetikhin–Turaev invariants, and conjecture that these invariants grow exponentially with growth rate given by the volume. Note that the TV and the WRT invariants grow only polynomially when we use usual roots of unity.

References

1. J.W. Alexander, A lemma on systems of knotted curves. Proc. Nat. Acad. Sci. USA **9**(3), 93–95 (1923)
2. J.E. Andersen, S.K. Hansen, Asymptotics of the quantum invariants for surgeries on the figure 8 knot. J. Knot Theory Ramif. **15**(4), 479–548 (2006). MR MR2221531
3. J.E. Andersen, R. Kashaev, A TQFT from quantum Teichmüller theory. Commun. Math. Phys. **330**(3), 887–934 (2014). MR 3227503
4. J.E. Andersen, R. Kashaev, Faddeev's quantum dilogarithm and state-integrals on shaped triangulations, in *Mathematical Aspects of Quantum Field Theories*, ed. by D. Calaque, T. Strobl. Mathematical Physics Studies (Springer, Cham, 2015), pp. 133–152. MR 3330241
5. J.E. Andersen, R.M. Kashaev, Quantum Teichmüller theory and TQFT, in *XVIIth International Congress on Mathematical Physics* (World Science Publications, Hackensack, 2014), pp. 684–692. MR 3204520
6. J.E. Andersen, S. Marzioni, Level *N* Teichmüller TQFT and complex Chern-Simons theory, in *Quantum Geometry of Moduli Spaces with Applications to TQFT and RNA Folding*, ed. by M. Schlichenmaier. Travaux Mathématiques, vol. 25. (Faculty of Science, Technology and Communication, University of Luxembourg, Luxembourg, 2017), pp. 97–146. MR 3700061
7. J.E. Andersen, J.-J.K. Nissen, Asymptotic aspects of the Teichmüller TQFT, in *Quantum Geometry of Moduli Spaces with Applications to TQFT and RNA Folding*, ed. by M. Schlichenmaier. Travaux Mathématiques, vol. 25. (Faculty of Science, Technology and Communication, University of Luxembourg, Luxembourg, 2017), pp. 41–95. MR 3700060
8. E. Artin, Theorie der Zöpfe. Abh. Math. Sem. Univ. Hamburg **4**(1), 47–72 (1925). MR 3069440
9. M. Atiyah, *The Geometry and Physics of Knots*. Lezioni Lincee. [Lincei Lectures] (Cambridge University Press, Cambridge, 1990). MR 1078014
10. R.J. Baxter, Partition function of the eight-vertex lattice model. Ann. Phys. **70**, 193–228 (1972). MR 0290733
11. R.J. Baxter, *Exactly Solved Models in Statistical Mechanics* (Academic, Inc. [Harcourt Brace Jovanovich, Publishers], London, 1989); Reprint of the 1982 original. MR 998375
12. J.S. Birman, *Braids, Links, and Mapping Class Groups* (Princeton University Press, Princeton, 1974). MR MR0375281 (51 #11477)
13. G. Burde, *Darstellungen von Knotengruppen*. Math. Ann. **173**, 24–33 (1967). MR MR0212787 (35 #3652)
14. G. Burde, H. Zieschang, M. Heusener, Knots, in *De Gruyter Studies in Mathematics*, vol. 5, extended edn. (De Gruyter, Berlin, 2014). MR 3156509
15. D. Calegari, Real places and torus bundles. Geom. Dedicata **118**, 209–227 (2006). MR 2239457

© The Author(s), under exclusive licence to Springer Nature Singapore Pte Ltd. 2018
H. Murakami, Y. Yokota, *Volume Conjecture for Knots*, SpringerBriefs in
Mathematical Physics 30, https://doi.org/10.1007/978-981-13-1150-5

16. Q. Chen, T. Yang, A volume conjecture for a family of Turaev-Viro type invariants of 3-manifolds with boundary (2015). arXiv:1503.02547
17. D. Cooper, M. Culler, H. Gillet, D.D. Long, P.B. Shalen, Plane curves associated to character varieties of 3-manifolds. Invent. Math. **118**(1), 47–84 (1994). MR MR1288467 (95g:57029)
18. G. de Rham, Introduction aux polynômes d'un nœud. Enseignement Math. **13**(2) (1967); 187–194 (1968). MR MR0240804 (39 #2149)
19. T. Dimofte, S. Gukov, Quantum field theory and the volume conjecture, in *Interactions Between Hyperbolic Geometry, Quantum Topology and Number Theory, Contemporary Mathematics*, ed. by A. Champanerkar, vol. 541 (Rhode Island American Mathematical Society, Providence, 2011), pp. 41–67. MR 2796627 (2012c:58037)
20. J. Dubois, Non abelian twisted Reidemeister torsion for fibered knots. Can. Math. Bull. **49**(1), 55–71 (2006). MR 2198719 (2006k:57064)
21. J. Dubois, V. Huynh, Y. Yamaguchi, Non-abelian Reidemeister torsion for twist knots. J. Knot Theory Ramif. **18**(3), 303–341 (2009). MR MR2514847
22. J. Dubois, R.M. Kashaev, On the asymptotic expansion of the colored Jones polynomial for torus knots. Math. Ann. **339**(4), 757–782 (2007). MR 2341899
23. L.D. Faddeev, Discrete Heisenberg-Weyl group and modular group. Lett. Math. Phys. **34**(3), 249–254 (1995). MR 1345554 (96i:46075)
24. R.H. Fox, Free differential calculus. I. Derivation in the free group ring. Ann. Math. **57**(2), 547–560 (1953). MR 0053938 (14,843d)
25. S. Garoufalidis, T.T.Q. Le, On the volume conjecture for small angles. arXiv: math.GT/0502163
26. S. Garoufalidis, T.T.Q. Lê, Asymptotics of the colored Jones function of a knot. Geom. Topol. **15**(4), 2135–2180 (2011). MR 2860990
27. S. Garoufalidis, I. Moffatt, D.P. Thurston, Non-peripheral ideal decompositions of alternating knots. arXiv:1610.09901
28. M. Gromov, *Volume and Bounded Cohomology*. Publications Mathématiques, vol. 56 (Institut des Hautes Études Scientifiques, Bures-sur-Yvette, 1982), pp. 5–99 (1983). MR 84h:53053
29. S. Gukov, H. Murakami, SL(2, ℂ) Chern-Simons theory and the asymptotic behavior of the colored Jones polynomial, in *Modular Forms and String Duality*, ed. by N. Yui, et al. Fields Institute Communications, vol. 54 (American Mathematical Society, Providence, 2008), pp. 261–277. MR 2454330
30. K. Habiro, On the colored Jones polynomials of some simple links. Sūrikaisekikenkyūsho Kōkyūroku **1172**, 34–43 (2000). MR 1 805 727
31. M. Heusener, An orientation for the SU(2)-representation space of knot groups, in *Proceedings of the Pacific Institute for the Mathematical Sciences Workshop "Invariants of Three-Manifolds"*, Calgary, 1999, vol. 127 (2003), pp. 175–197. MR 1953326 (2003m:57013)
32. K. Hikami, Quantum invariant for torus link and modular forms. Commun. Math. Phys. **246**(2), 403–426 (2004). MR 2 048 564
33. K. Hikami, H. Murakami, Colored Jones polynomials with polynomial growth. Commun. Contemp. Math. **10**(suppl. 1), 815–834 (2008). MR MR2468365
34. C.D. Hodgson, S.P. Kerckhoff, Rigidity of hyperbolic cone-manifolds and hyperbolic Dehn surgery. J. Differ. Geom. **48**(1), 1–59 (1998). MR MR1622600 (99b:57030)
35. W.H. Jaco, P.B. Shalen, Seifert fibered spaces in 3-manifolds. Mem. Am. Math. Soc. **21**(220), viii+192 (1979). MR 81c:57010
36. K. Johannson, *Homotopy Equivalences of 3-Manifolds with Boundaries*. Lecture Notes in Mathematics, vol. 761 (Springer, Berlin, 1979). MR 82c:57005
37. V.F.R. Jones, A polynomial invariant for knots via von Neumann algebras. Bull. Am. Math. Soc. (N.S.) **12**(1), 103–111 (1985). MR 86e:57006
38. R.M. Kashaev, A link invariant from quantum dilogarithm. Mod. Phys. Lett. A **10**(19), 1409–1418 (1995). MR 1341338

39. R.M. Kashaev, The hyperbolic volume of knots from the quantum dilogarithm. Lett. Math. Phys. **39**(3), 269–275 (1997). MR 1434238
40. R.M. Kashaev, O. Tirkkonen, A proof of the volume conjecture on torus knots, Zap. Nauchn. Sem. S.-Peterburg. Otdel. Mat. Inst. Steklov. (POMI) **269** (2000); no. Vopr. Kvant. Teor. Polya i Stat. Fiz. 16, 262–268, 370. MR 1805865
41. R.M. Kashaev, Y. Yokota, On the volume conjecture for 5_2 (Preprint, 2012)
42. C. Kassel, V. Turaev, Braid groups, in *Graduate Texts in Mathematics*, vol. 247 (Springer, New York, 2008); With the graphical assistance of Olivier Dodane. MR 2435235
43. L.H. Kauffman, *On Knots*. Annals of Mathematics Studies, vol. 115 (Princeton University Press, Princeton, 1987). MR MR907872 (89c:57005)
44. L.H. Kauffman, State models and the Jones polynomial. Topology **26**(3), 395–407 (1987). MR 88f:57006
45. R. Kirby, P. Melvin, The 3-manifold invariants of Witten and Reshetikhin-Turaev for sl(2, **C**). Invent. Math. **105**(3), 473–545 (1991). MR 92e:57011
46. A.N. Kirillov, N. Yu. Reshetikhin, Representations of the algebra U_q(sl(2)), q-orthogonal polynomials and invariants of links, in *Infinite-Dimensional Lie Algebras and Groups*, Luminy-Marseille, 1988, ed. by V.G. Kac. Advanced Series in Mathematical Physics, vol. 7 (World Science Publishing, Teaneck, 1989), pp. 285–339. MR MR1026957 (90m:17022)
47. P. Kirk, E. Klassen, Chern-Simons invariants of 3-manifolds decomposed along tori and the circle bundle over the representation space of T^2. Commun. Math. Phys. **153**(3), 521–557 (1993). MR 94d:57042
48. P. Kirk, C. Livingston, Twisted Alexander invariants, Reidemeister torsion, and Casson-Gordon invariants. Topology **38**(3), 635–661 (1999). MR 1670420 (2000c:57010)
49. T. Kitano, Twisted Alexander polynomial and Reidemeister torsion. Pac. J. Math. **174**(2), 431–442 (1996). MR 1405595 (97g:57007)
50. T. Kohno, *Conformal Field Theory and Topology*. Translations of Mathematical Monographs, vol. 210 (American Mathematical Society, Providence, 2002); Translated from the 1998 Japanese original by the author, Iwanami Series in Modern Mathematics. MR 1905659
51. T.T.Q. Le, Varieties of representations and their subvarieties of cohomology jumps for knot groups. Mat. Sb. **184**(2) 57–82 (1993). MR 1214944
52. T.T.Q. Le, A.T. Tran, On the volume conjecture for cables of knots. J. Knot Theory Ramif. **19**(12), 1673–1691 (2010). MR 2755495
53. W.B.R. Lickorish, *An Introduction to Knot Theory*. Graduate Texts in Mathematics, vol. 175 (Springer, New York, 1997). MR 98f:57015
54. A. Markoff, Über die freie Äquivalenz der geschlossenen Zöpfe. Rec. Math. Moscou, n. Ser. **1**, 73–78 (1936, German)
55. J.E. Marsden, M.J. Hoffman, *Basic Complex Analysis* (W. H. Freeman and Company, New York, 1987). MR 88m:30001
56. G. Masbaum, P. Vogel, 3-valent graphs and the Kauffman bracket. Pac. J. Math. **164**(2), 361–381 (1994). MR MR1272656 (95e:57003)
57. C.T. McMullen, *The Evolution of Geometric Structures on 3-Manifolds*. The Poincaré Conjecture, Clay Mathematics Proceedings, vol. 19 (American Mathematical Society, Providence, 2014), pp. 31–46. MR 3308757
58. R. Meyerhoff, Density of the Chern-Simons invariant for hyperbolic 3-manifolds, in *Low-Dimensional Topology and Kleinian Groups*, Coventry/Durham, 1984, ed. by D.B.A Epstein. London Mathematical Society Lecture Note Series, vol. 112 (Cambridge University Press, Cambridge, 1986), pp. 217–239. MR 903867
59. J. Milnor, Whitehead torsion. Bull. Am. Math. Soc. **72**, 358–426 (1966). MR MR0196736 (33 #4922)
60. J. Milnor, Hyperbolic geometry: the first 150 years. Bull. Am. Math. Soc. (N.S.) **6**(1), 9–24 (1982). MR 82m:57005

61. H.R. Morton, The coloured Jones function and Alexander polynomial for torus knots. Math. Proc. Camb. Philos. Soc. **117**(1), 129–135 (1995). MR 1297899 (95h:57008)
62. H. Murakami, Some limits of the colored Jones polynomials of the figure-eight knot. Kyungpook Math. J. **44**(3), 369–383 (2004). MR MR2095421
63. H. Murakami, The colored Jones polynomials and the Alexander polynomial of the figure-eight knot. JP J. Geom. Topol. **7**(2), 249–269 (2007). MR MR2349300 (2008g:57014)
64. H. Murakami, The coloured Jones polynomial, the Chern-Simons invariant, and the Reidemeister torsion of the figure-eight knot. J. Topol. **6**(1), 193–216 (2013). MR 3029425
65. H. Murakami, The twisted Reidemeister torsion of an iterated torus knot (2016). arXiv: 1602.04547
66. H. Murakami, J. Murakami, The colored Jones polynomials and the simplicial volume of a knot. Acta Math. **186**(1), 85–104 (2001). MR 1828373
67. H. Murakami, J. Murakami, M. Okamoto, T. Takata, Y. Yokota, Kashaev's conjecture and the Chern-Simons invariants of knots and links. Exp. Math. **11**(3), 427–435 (2002). MR 1 959 752
68. H. Murakami, Y. Yokota, The colored Jones polynomials of the figure-eight knot and its Dehn surgery spaces. J. Reine Angew. Math. **607**, 47–68 (2007). MR MR2338120
69. W.D. Neumann, Extended Bloch group and the Cheeger-Chern-Simons class. Geom. Topol. **8**, 413–474 (2004) (electronic). MR MR2033484 (2005e:57042)
70. W.D. Neumann, D. Zagier, Volumes of hyperbolic three-manifolds. Topology **24**(3), 307–332 (1985). MR 87j:57008
71. T. Ohtsuki, On the asymptotic expansion of the Kashaev invariant of the 5_2 knot. Quantum Topol. **7**(4), 669–735 (2016). MR 3593566
72. T. Ohtsuki, Y. Yokota, On the asymptoitc expansions of the Kashaev invariant of the knots with 6 crossings, Math. Proc. Camb. Philos. Soc. (To appear). https://doi.org/10.1017/S0305004117000494
73. J. Porti, Torsion de Reidemeister pour les variétés hyperboliques. Mem. Am. Math. Soc. **128**(612), x+139 (1997). MR MR1396960 (98g:57034)
74. K. Reidemeister, *Knotentheorie* (Springer, Berlin/New York, 1974), Reprint. MR 0345089
75. R. Riley, Nonabelian representations of 2-bridge knot groups. Q. J. Math. Oxf. Ser. (2) **35**(138), 191–208 (1984). MR MR745421 (85i:20043)
76. D. Rolfsen, *Knots and Links*. Mathematics Lecture Series, vol. 7 (Publish or Perish, Inc., Houston, 1990), Corrected reprint of the 1976 original. MR 1277811 (95c:57018)
77. M. Rosso, V. Jones, On the invariants of torus knots derived from quantum groups. J. Knot Theory Ramif. **2**(1), 97–112 (1993). MR 1209320 (94a:57019)
78. M. Sakuma, Y. Yokota, An application of non-positively curved cubings of alternating links. Proc. Am. Math. Soc. 146, 3167–3178 (2018)
79. D. Thurston, *Hyperbolic Volume and the Jones Polynomial*. Lecture notes, École d'été de Mathématiques 'Invariants de nœuds et de variétés de dimension 3', Institut Fourier – UMR 5582 du CNRS et de l'UJF Grenoble (France) du 21 juin au 9 juillet 1999. http://www.math.columbia.edu/~dpt/speaking/Grenoble.pdf
80. W.P. Thurston, *The Geometry and Topology of Three-Manifolds*, Electronic version 1.1 – Mar 2002. http://www.msri.org/publications/books/gt3m/
81. W.P. Thurston, *Three-Dimensional Geometry and Topology, Volume 1*, ed. by S. Levy. Princeton Mathematical Series, vol. 35 (Princeton University Press, Princeton, 1997). MR 97m:57016
82. V. Turaev, *Introduction to Combinatorial Torsions*. Lectures in Mathematics ETH Zürich (Birkhäuser Verlag, Basel, 2001), Notes taken by Felix Schlenk. MR 1809561 (2001m:57042)
83. V.G. Turaev, The Yang-Baxter equation and invariants of links. Invent. Math. **92**(3), 527–553 (1988). MR 939474
84. R. van der Veen, Proof of the volume conjecture for Whitehead chains. Acta Math. Vietnam. **33**(3), 421–431 (2008). MR MR2501851
85. F. Waldhausen, Algebraic K-theory of generalized free products. I, II, Ann. Math. (2) **108**(1), 135–204 (1978). MR 0498807 (58 #16845a)

86. J. Weeks, Computation of hyperbolic structures in knot theory, in *Handbook of Knot Theory*, ed. by W.W. Menasco (Elsevier B. V., Amsterdam, 2005), pp. 461–480. MR 2179268
87. H. Wenzl, On sequences of projections. C. R. Math. Rep. Acad. Sci. Can. **9**(1), 5–9 (1987). MR 88k:46070
88. E. Witten, Quantum field theory and the Jones polynomial. Commun. Math. Phys. **121**(3), 351–399 (1989). MR 990772
89. Wolfram Research, Inc., Mathematica, version 11.1 (2017)
90. Y. Yamaguchi, A relationship between the non-acyclic Reidemeister torsion and a zero of the acyclic Reidemeister torsion. Ann. Inst. Fourier (Grenoble) **58**(1), 337–362 (2008). MR 2401224 (2009c:57039)
91. C.N. Yang, Some exact results for the many-body problem in one dimension with repulsive delta-function interaction. Phys. Rev. Lett. **19**, 1312–1315 (1967). MR 0261870
92. Y. Yokota, On the volume conjecture for hyperbolic knots. arXiv: math.QA/0009165
93. Y. Yokota, From the Jones polynomial to the A-polynomial of hyperbolic knots. Interdiscip. Inf. Sci. **9**(1), 11–21 (2003). MR MR2023102 (2004j:57014)
94. Y. Yokota, On the complex volume of hyperbolic knots. J. Knot Theory Ramif. **20**(7), 955–976 (2011). MR 2819177
95. T. Yoshida, The η-invariant of hyperbolic 3-manifolds. Invent. Math. **81**(3), 473–514 (1985). MR 87f:58153
96. H. Zheng, Proof of the volume conjecture for Whitehead doubles of a family of torus knots. Chin. Ann. Math. Ser. B **28**(4), 375–388 (2007). MR MR2348452
97. C.K. Zickert, The volume and Chern-Simons invariant of a representation. Duke Math. J. **150**(3), 489–532 (2009). MR 2582103

Index

© The Author(s), under exclusive licence to Springer Nature Singapore Pte Ltd. 2018   119
H. Murakami, Y. Yokota, *Volume Conjecture for Knots*, SpringerBriefs in
Mathematical Physics 30, https://doi.org/10.1007/978-981-13-1150-5

Printed in the United States
By Bookmasters